Stimuli-Responsive Polymeric Films and Coatings

ACS SYMPOSIUM SERIES **912**

Stimuli-Responsive Polymeric Films and Coatings

Marek W. Urban, Editor
The University of Southern Mississippi

Sponsored by the
**ACS Division of Polymeric Materials:
Science and Engineering, Inc.**

American Chemical Society, Washington, DC

Library of Congress Cataloging-in-Publication Data

Stimuli-responsive polymeric films and coatings / Marek W. Urban, editor ; sponsored by the ACS Division of Polymeric Materials: Science and Engineering, Inc.

p. cm.—(ACS symposium series ; 912)

"Developed from a symposium sponsored by the Division of Polymeric Materials: Science and Engineering, Inc. at the 227th National Meeting of the American Chemical Society, Anaheim, California, March 28–Aapril 1, 2004"—CIP t.p. verso.

Includes bibliographical references and index.

ISBN 0-8412-3932-0 (alk. paper)

1. Polymers—Congresses. 2. Thin films—Congresses. 3. Coatings—Congresses.

I. Urban, Marek W., 1953- II. American Chemical Society. Division of Polymeric Materials: Science and Engineering, Inc. III. American Chemical Society. Meeting (227th : 2004 : Anaheim, Calif.) IV. Series.

QD382.S75S75   2005
547'.7—dc22   2005042134

The paper used in this publication meets the minimum requirements of American National Standard for Information Sciences—Permanence of Paper for Printed Library Materials, ANSI Z39.48–1984.

Copyright © 2005 American Chemical Society

Distributed by Oxford University Press

All Rights Reserved. Reprographic copying beyond that permitted by Sections 107 or 108 of the U.S. Copyright Act is allowed for internal use only, provided that a per-chapter fee of $30.00 plus $0.75 per page is paid to the Copyright Clearance Center, Inc., 222 Rosewood Drive, Danvers, MA 01923, USA. Republication or reproduction for sale of pages in this book is permitted only under license from ACS. Direct these and other permission requests to ACS Copyright Office, Publications Division, 1155 16th Street, N.W., Washington, DC 20036.

The citation of trade names and/or names of manufacturers in this publication is not to be construed as an endorsement or as approval by ACS of the commercial products or services referenced herein; nor should the mere reference herein to any drawing, specification, chemical process, or other data be regarded as a license or as a conveyance of any right or permission to the holder, reader, or any other person or corporation, to manufacture, reproduce, use, or sell any patented invention or copyrighted work that may in any way be related thereto. Registered names, trademarks, etc., used in this publication, even without specific indication thereof, are not to be considered unprotected by law.

PRINTED IN THE UNITED STATES OF AMERICA

Foreword

The ACS Symposium Series was first published in 1974 to provide a mechanism for publishing symposia quickly in book form. The purpose of the series is to publish timely, comprehensive books developed from ACS sponsored symposia based on current scientific research. Occasionally, books are developed from symposia sponsored by other organizations when the topic is of keen interest to the chemistry audience.

Before agreeing to publish a book, the proposed table of contents is reviewed for appropriate and comprehensive coverage and for interest to the audience. Some papers may be excluded to better focus the book; others may be added to provide comprehensiveness. When appropriate, overview or introductory chapters are added. Drafts of chapters are peer-reviewed prior to final acceptance or rejection, and manuscripts are prepared in camera-ready format.

As a rule, only original research papers and original review papers are included in the volumes. Verbatim reproductions of previously published papers are not accepted.

ACS Books Department

Contents

Preface..ix

1. Stimuli-Responsive Macromolecules and Polymeric Coatings...............1
   Anna M. Urban and Marek W. Urban

Synthetic Aspects of Stimuli-Responsive Polymers

2. Controlling Polymer Chain Topology and Architecture
   by ATRP from Flat Surfaces..28
   Joanna Pietrasik, Lindsay Bombalski, Brian Cusick, Jinyu Huang,
   Jeffrey Pyun, Tomasz Kowalewski, and Krzysztof Matyjaszewski

3. Synthesis of Terminally Functionalized (Co)Polymers via
   Reversible Addition Fragmentation Chain Transfer
   Polymerization and Subsequent Immobilization to Solid
   Surfaces with Potential Biosensor Applications....................................43
   Charles W. Scales, Anthony J. Convertine, Brent S. Sumerlin,
   Andrew B. Lowe, and Charles L. McCormick

4. Synthesis and Application of Polyelectrolyte Brushes........................55
   Stephen G. Boyes, Crystal Cyrus, Bulent Akgun, Adam Caplan,
   Brian Mirous, and William J. Brittain

5. Gradient Stimuli-Responsive Polymer Grafted Layers.......................68
   Sergiy Minko, Leonid Ionov, Alexander Sydorenko,
   Nikolay Houbenov, Manfred Stamm, Bogdan Zdyrko,
   Viktor Klep, and Igor Luzinov

6. Tailoring of Thin Polymer Films Chemisorbed onto Conductive
   Surfaces by Electrografting...84
   C. Jérôme and R. Jérôme

Physico-Chemical Aspects of Stimuli-Responsive Polymers

7. Temperature-Responsive Films of Amphiphilic Poly(*N*-isopropylacrylamides) of Various Architectures at the Air–Water Interface..................107
   Roger C. W. Liu and Françoise M. Winnik

8. Stimuli-Responsive Behavior of Sodium Dodecyl Sulfate and Sodium Dioctyl Sulfosuccinate during Coalescence of Colloidal Particles..................122
   David J. Lestage, W. Reid Dreher, and Marek W. Urban

9. Fluorescence Methods for Latex Film Formation..................137
   Önder Pekcan and Ertan Arda

10. Finite Element Modeling for Scratch Damage of Polymers..................166
    G. T. Lim, J. N. Reddy, and H.-J. Sue

11. An Artificial Neural Network Approach to Stimuli-Response Mechanisms in Complex Systems..................181
    Eduardo Nahmad-Achar and Javier E. Vitela

12. Maleimide–Vinyl Ether-Based Polymer Dispersed Liquid Crystals..................195
    Molly L. Hladik, Askym F. Senyurt, Charles E. Hoyle, Joe B. Whitehead, Charles M. Werneth, and Garfield T. Warren

13. Stimuli-Responsive Supramolecular Solids: Functional Porous and Inclusion Materials..................214
    Dmitriy V. Soldatov

Indexes

Author Index..................235

Subject Index..................237

Preface

Materials' response to internal or external stimuli formulates the basis of novel materials of the 21st Century. Polymeric materials, particularly polymeric films, have paced the evolution of technologies for the past few decades and their importance in the advancement of global growth is indispensable. Ranging from applications as protective or beatifying coatings applied to a variety of surfaces, to biofilms or medical devices saving human lives, or conductive polymers and insulators for electronic devices, polymeric films have been of importance due to their versatility of applications. These developments were possible due to tremendous scientific advances during the past few decades. Although scattered research studies indicated that the distribution of molecular entities within a given film may lead to new and often puzzling surface and interfacial properties, it was only during the past decade that the importance of stratification processes in polymeric films was recognized. This is particularly important in view of the fact that such processes may generate response films by the deliberate use of cognitively driven molecular entities, thus generating desirable stimuli-responsive interactions. One well-recognized stimuli-responsive example is the concepts of memory shapes known for many years in metals, but recent developments in nanocomposites offer new opportunities for polymeric films and coatings. One can envision that incorporation of carbon nanotubes into a polymer film will allow recoveries of the original shape when exposed to a stimulus such as electricity or heat. There are other examples and all demonstrate that it is enormously important to understand atomic and molecular levels interactions in polymeric films in order to gain knowledge that is the key enabler to many next-generation devices. These tasks represent extreme scientific and technological challenges and hopefully this monograph will take us one step closer to the development of new and often unprecedented stimuli-responsive characteristics of polymeric films and coatings.

While the introductory chapter provides a brief overview of a variety of stimuli-responsive behaviors in polymers, the first section of this book is devoted to synthetic and mechanistic aspects of polymers that are capable of controlling topologies and architectures of surfaces as well as synthetic immobilization of polymers on surfaces. Synthesis and applications of polyelectrolyte brushes as well as gradient stimuli-responsive layers are discussed along with chemisorption processes on conductive surfaces. The second section of this monograph deals with physicochemical aspects of stimuli-responsive polymers with the focus of conformational changes resulting from temperature, stratification of individual components as a result of external or internal stimuli, film formation from colloidal solutions, liquid crystalline films, mechanical damage modeling and interfacial stresses, neural elements modeling of coatings, and supra-molecular stimuli-responsive pores and inclusions.

One of the best learning and at the same time puzzling platforms of stimuli-responsiveness are biological systems. We hope this book will provide a starting point for this process and will encourage further studies to understand responses in polymeric films via interdisciplinary research activities. Using this knowledge, the development of synthetic routes for macromolecular chain responses and stimuli functions enabling new technologies and predicting environmental responses will become possible.

Marek W. Urban
School of Polymers and High Performance Materials
The University of Southern Mississippi
118 College Avenue
Hattiesburg, MS 39406

Chapter 1

Stimuli-Responsive Macromolecules and Polymeric Coatings

Anna M. Urban[1,2] and Marek W. Urban[1,*]

[1]School of Polymers and High Performance Materials, Shelby F. Thames Polymer Science Research Center, Department of Polymer Science, The University of Southern Mississippi, Hattiesburg, MS 39406
[2]Current addresses: Department of Medicinal Chemistry, University of Minnesota, Minneapolis, MN 55455
*Corresponding author: marek.urban@usm.edu

Polymers represent a unique class of materials which are capable of forming architectural arrangements at molecular, nano- and higher levels of organization. While numerous properties made polymers highly attractive materials in many applications, their stimuli-responsive characteristics have been recognized only recently. This chapter provides an overview of a variety of responses of polymers and polymeric films that exhibit pro-active behaviors in response to internal or external stimuli. These "smart" macromolecules offer a number of unique attributes with potential emerging applications in medicine, pharmacology, nanotechnology, biomaterials, or other fields of materials science and materials engineering. The stimuli-responsive characteristics described herein focus on physical and chemical stimuli that result in numerous responses leading to unique behaviors in complex environments.

Introduction

Although polymeric materials have paced the evolution of technologies for over 50 years, their importance in the development of environmental stimuli functions has been recognized only recently. This chapter provides an overview of past and recent advances in the development and analysis of stimuli-responsive polymeric materials which play a key driving role in advancing polymer-based biomaterials, nano-technologies and materials science and engineering. Understanding responses of biological systems formulates a basis for the development of synthetic routes leading to the developments of responsive macromolecules with stimuli-responsive functions, thus paving opportunities for novel technologies utilizing smart polymeric materials. Thus, the main challenges are molecular designs containing environmentally compliant macromolecules that function in complex environments, and processes leading to their film formation. Since stimuli-responsiveness of man-made materials is clearly demonstrated in the development of biomaterials which are typically derived from biological building blocks, such building blocks as aminoacids play an enormous role in the development of medical devices such as polymer-coated stents,[1] or drug delivery systems,[2] which are probably one of the most illustrative examples of the recent utilization of stimuli-responsive polymers.

The vast majority of recognition and response driven systems are initiated at or near surfaces. Examples of such processes are bio-absorption or bio-adsorption. For that reason significant research efforts are focused on synthetic aspects of pro-active macromolecules which are initiated near the areas of excess of interfacial energy. Therefore, significant efforts are centered on materials design and engineering processing that focus on spatial compositional and structural gradients near the surfaces or interfaces. These characteristics are of importance in biomaterials and biomechanics, fluidics, tribology, geology, and nanotechnology, just to name a few. Since chemical and physical designs of surfaces play the leading role in generation of successful response-driven films, polymer synthesis, formulation and interactions of individual components, and processes involved in their film formation are among the main elements necessary to mimic biological systems.

Stimuli-responsive polymers and polymeric films along with the processes associated with their formation are clearly distinguishable from their traditional functional characteristics. While chemical functional groups in a given spatial physico-chemical environment determine polymer functional properties, stimuli-responsive polymers are active and capable of dynamic responses to internal or external stimuli. These behaviors make the development of polymeric films most challenging. Thus, one aspect is the design of responsive macromolecular chains, and another is generating responsive surface properties as a result of film formation. While the former focuses on the responses in solution environments, the latter preferably occurs during or after solidification across the film thickness or near surfaces or interfaces. Figure 1 schematically illustrates the distinction between functionality and stimuli-responsiveness which may result from chemical structures or physical film formation resulting in arrays that are stimuli-responsive.

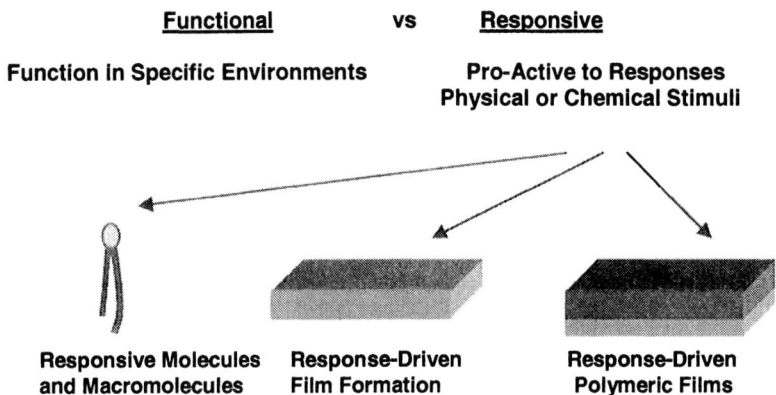

Figure 1. Pictorial representation of functional and responsive macromolecules and polymeric films.

The following sections outline selected physical and chemical examples of stimuli-responsive behaviors of polymeric materials that are significant for the current and future developments or "smart" materials for bio- and nanotechnologies that exhibit responses to mechanical stresses, temperature, pH, electric and magnetic fields, and other existing and emerging fields of science and engineering.

Stress-Responses in Polymers

It is well established that the uniaxial elongation of a polymer network containing randomly oriented crystalline and amorphous components typically leads to the crystalline phase orientation parallel to the strain axis. Consequently, polymers may experience more oriented morphologies due to elongation, and exhibit either diminished or enhanced rates of diffusion.[3,4] An increase in ordering as a result of elongation thus allows for closer spatial packing of chains and crystallites, which, in turn, reduces the number of diffusion pathways and hence slows down the diffusion process. This is the most commonly known physical stimulus where a simple elongation process may often exhibit shape or thermal history memories.

Thermoplastic polymers have been known to respond to external mechanical elongations and many polymers that are hybrids of amorphous and crystalline components exhibit crystallographic changes resulting from elongations. For example, the crystal phase transformation in poly(butylene terephthalate) (PBT) [5,6,7] and poly(ethylene terephthalate) (PET) upon application of external stresses has been examined on numerous occasions and showed that the phase transformation may or may not be reversible, and depends on many factors.

This behavior, however, may not necessarily be common for all semicrystalline polymers that undergo uniaxial elongation. For example, poly(tetrafluoroethylene) (PTFE) exhibits an unusual thickening in the direction perpendicular to the strain upon elongation. This effect is referred to as a negative Poisson ratio (NPR), and has been seen in pre-rolled PTFE with up to a 12% strain.[8,9] As shown in Figure 2, rotation of anisotropic crystals in a film in the director perpendicular to its elongation has been attributed to this behavior.[10] The film thickens in the range of 4-12 % elongation (Figure 1, b) resulting in expansion in the direction perpendicular to the elongation direction and further elongation results in the film thinning (Figure 1, c and d). It has been also shown that realignment of crystallites alters diffusion rates through polymers.[11]

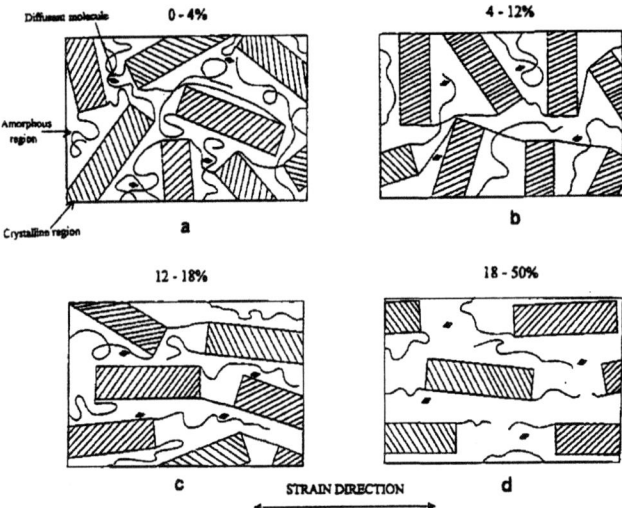

Figure 2. Morphological changes in PTFE as a function of elongation. (a) 0 to 4% strain; (b) 4 to 12% strain; (c) 12 to 18% strain; (d) 18 to 50% strain. Permeability increases and microvoids are formed as a result of crystalline slippage and chain rupture (Reproduced with permission from reference 10. Copyright 1994 Elsevier.)

In essence, viscoelastic responses to external stresses can be driven by several factors and the properties and alignment of crystalline and amorphous regions that often result from thermal history appear to be one of the key components. One area that is of particular interest is the development of thermoplastic systems with memory shapes. A general concept in creating these materials is to achieve phase segregation either via the presence of two phases with different glass or melting transition temperatures. While the phase that is responsible for the permanent shape exhibits higher transition temperatures, the lower transition phase acts as a switch responding to external stimuli. For example, it is possible to design polyethylene tubes that upon exposure to 37°C will coil up, thus opening various possibilities in being utilized in biomedical divices, or materials based on biocompatible ε-caprolactone and p-dioxanone natural building blocks.[12] In thermosetting polymers the content of elastic and viscous components and their miscibility or compatibility in each other will dictate their responsiveness to external stimuli and these properties are typically reflected in the storage and loss moduli as a function of temperature.

Using the shape memory materials, minimally invasive surgery[13] has made it possible by implantation of small medical devices with laprascopes. Thus, developments of biocompatible shape-memory polymers which are able to memorize a permanent shape that differs from the initial temporary shape, could potentially allow devices to be introduced into the body in a compressed temporary shape. Such implants would then expand on demand to their permanent shape.[14] This thermally induced shape-memory effect has been described for various classes of materials which are nondegradable in physiological environments; for example, polymers[15,16] such as polyurethanes,[17,18,19] poly(styrene-block-butadiene),[20] and polynorbornene.[21,22] Other such materials include hydrogels,[23,24] metallic alloys,[25] and ceramics.[26] However, many of these materials lack biocompatibility or compliance in mechanical properties. Subsequently, multi-blocked copolymers have been developed that can be elongated up to 1000% before breaking. Deformations between permanent and temporary shape up to 400% are feasible, and compared to a maximum 8% deformation for Ni-Ti-alloys, polymeric films exhibit highly promising properties.

Highly cross-linked polymers have been used as matrices for structural adhesives, foamed structures, insulators for electronics, and numerous other applications.[27,28] Mechanical properties such as high modulus, high fracture strength, and solvent resistance stem from these densely cross-linked structures. Although useful, these materials are irreversibly damaged by high stresses[29,30] due to the formation of cracks and this damage may lead to loss in the load-carrying capacity of polymeric structural engineering materials.[31,32] Therefore, the re-mending ability of polymeric materials has been developed, particularly the development of a rigid solid transparent organic polymeric material which is able to restore fractured parts multiple times without the use of catalysts or special surface treatments.[33] A quest is still open on the development of completely self-heeling polymeric coatings that would repair themselves upon external damage.

Although there have been numerous static and dynamic rheological studies that correlate solution or melt viscosities to polymer molecular weight and molecular weight distribution, concentration levels, or temperature, limited studies exist concerning stimuli-responsiveness with respect to the effect of molecular and structural effects of polymers. Although there are numerous examples that focus on practical measurements of solution viscosity and its responses to dynamic and/or static directional or randomly applied shear stresses, correlations between polymer structural features and rheological behavior are often lacking.[34] Thus, there are numerous opportunities for advancing limited knowledge as polymer configurations in solution are influenced by flowing fluids, and forces exerted back on the fluid are directly related to the molecular conformations. For example, flows with large rotational components do not generally perturb polymer configurations substantially, while flows with dominant extensional components may substantially alter polymer orientation and stretch.[35] Since extensional flows are essential in many polymer-processing applications such as coatings, injection molding, and fiber drawing,[36] future studies on rheological stimuli-responsive behaviors will have many scientific and technological implications.

Temperature and pH

Polymer systems that undergo a conformational change or phase transition in response to an external stimulus are probably one of the most suited examples of the stimuli-responsive behavior. Taking this concept one step further and mimicking the external stimuli functions of biological systems, and translating them into synthetic macromolecules which are responsive to temperature, electrolyte, or changes in pH or light, opens up a number of opportunities for the development of new materials. Such materials are of great scientific and technological importance, as they may serve in many practical and diverse applications which may vary from water treatment, enhanced oil recovery processes, controlled drug release, or other emerging biotechnologies.

In spite of low thermal expansion coefficient and conductivities, the most commonly utilized stimulus is temperature. For example, it is well known that one of the unique properties of linear poly-n-isopropylacrylamide (PNIPAm) is its lower critical solution temperature (LCST) of 32°C in water,[37] where microgel PNIPAm particles exhibit a conformational transition. The uniqueness of this behavior shown in Figure 3 comes from the fact that below the LCST the polymer expands, becoming hydrophilic and water soluble, while above the LCST the polymer shrinks, becoming hydrophobic and water insoluble. In these PNIPAm microgels, hydrogen bonding plays the role of increasing chain interaction through polymer-solvent interactions, thus the noted increased conformational temperature change from the linear LCST.[37] However, which portion of the polymer, the backbone or its side chains, responsible for its behavior is not fully understood. PNIPAm-based microgels may serve as temperature-dependent protein binding sites or recyclable sequestrators for heavy metals.[38,39] Detailed temperature-dependent rheological studies have also been reported.[40]

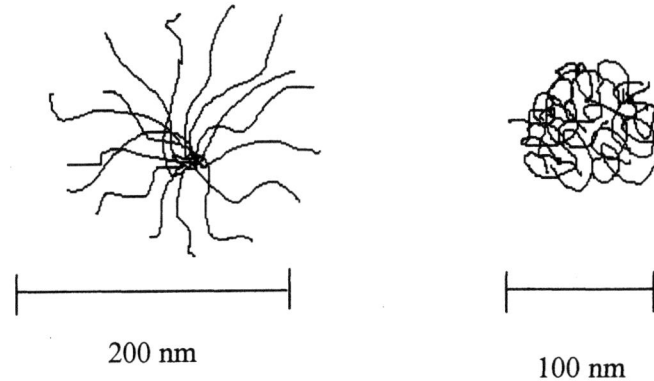

Figure 3. Size changes in PNIPAm resulting from temperatures changes.

Practical importance of pH-responsive gels vary from alkali-swellable latexes based on methacrylic acid used as thickeners for cosmetic and pharmaceuticals[41] to pH-induced swellable poly(methyl methacrylate-*stat*-methacrylic acid) latexes that deswell by alcohols and poly(ethylene oxide).[42] Along these lines, the synthesis of acid-swellable microgels of monodisperse poly(4-vinylpyridine)-based (P4VP) latex and the production of robust, ordered microgel films by chemical cross-linking via quaternization was also reported, [43] with applications as a new stationary phase for liquid chromatography.[44] Poly-(2-vinylpyridine)-based (P2VP) microgels[45] and "nanogels"[46] that comprise both thermo-responsive poly-NIPAm cores and pH-responsive P2VP-based coronas as well as acrylamide-based microgels via inverse emulsion polymerization were also reported.[47] More recently, the synthesis and characterization of a new class of acid-swellable, sterically stabilized microgel particles based on 2-(diethylamino)ethyl methacrylate (DEA) or 2-(diisopropylamino)ethyl methacrylate (DPA) were reported which swell at physiological pH (7.4), thus having potential biomedical applications.[48]

Recent studies[49] have shown that combining pH-responsive poly(acrylic acid) (PAA) and temperature-responsive PNIPAAm creates systems that respond to combined external stimuli. This was accomplished via reversible addition-fragmentation chain transfer (RAFT) copolymerization of respective monomers in the presence of a RAFT agent, and Figure 4 illustrates possible modes of aggregate formation with pH and temperature dependence. A combination of pH- and thermo-responses such as that shown in Figure 4 may have numerous applications in drug or protein research.

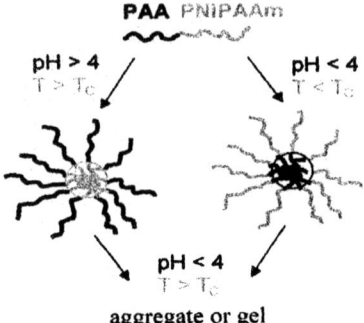

Figure 4. Proposed modes of aggreagate formation for PNIPAAm-b-PAA in aqueous solution with pH and temperature dependence (Reproduced from reference 35. Copyright 2003 American Chemical Society.)

Another building unit is 2-hydroxyethyl methacrylate (HEMA) which is a widely recognized monomer used in the manufacturing process of soft contact and intraocular lenses.[50,51,52] Due to excellent biocompatibility and good blood compatibility HEMA copolymers are of particular interest,[53,54] and recent studies have shown the synthesis of several novel HEMA-based block copolymers in which the HEMA block was either thermo-responsive or permanently hydrophilic, depending on its degree of polymerization, the pH of the solution, and the nature of the copolymer block.[55] As shown in Figure 5, self-assembly of the tri-block polyethylene oxide-2-hydroxyethyl methacrylate-2-(diethylamino) ethyl methacrylate (PEO-HEMA-DEA) occurs upon heating-cooling cycle. Below 7 °C, HEMA blocks form the hydrophilic inner shell of the micelle corona, but above that temperature, the same HEMA blocks are hydrophobic outer layers with the DEA core.

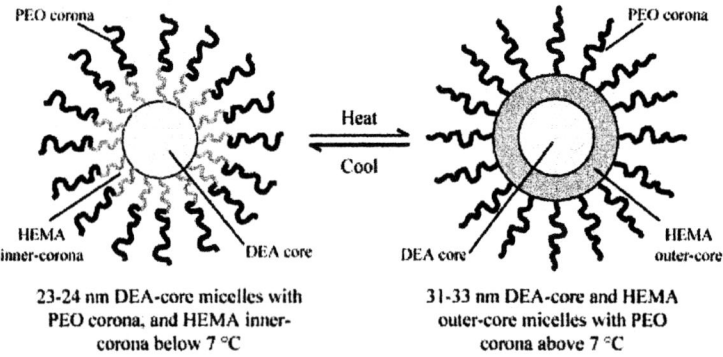

Figure 5. Temperature dependent self-assembly of a 0.50% w/v solution of a PEO45-HEMA30-DEA50 triblock copolymer at pH 11 (Reproduced from reference 35. Copyright 2003 American Chemical Society.)

As mentioned earlier, the presence of suitable chemical entities spatially placed in a given environment typically generates stimuli-responsiveness in biological systems. Amino acids, hydroxyls, phosphate esters, acetyls, thiols, or carboxyl-terminated species are just a few examples generating various forms of responsiveness in nature, either by chemical reactions with other groups, or hydrophilic and/or hydrophobic interactions. Polyvinylamine (p-VAm) as well as its derivatives are important water-soluble synthetic polymers due to various structural modification possibilities and functional properties. P-VAm, a linear polymer with all primary amine groups directly bonded to the main chain, is polycationic in nature and may form chelating complexes with certain heavy metal ions.[56,57] It may also act as a flocculant[58] or a support matrix to immobilize enzymes,[59] and modified p-VAm may serve as a polymeric surfactant[60] capable of emulating natural enzymes.[61] Along the same lines, n-vinylformamide (NVF) and n-vinylisobutyramide (NVIBA), and their derivates with primary amino groups, have been synthesized into thermo- and pH-responsive copolymers. For example, the hydrolysis of poly(NVF-co-NVIBA) hydrogels (NVF content is about 30mol%) can be carried out under acidic conditions and such products exhibit pH and/or thermal responsiveness. Figure 6 illustrates dimensional changes resulting from the interplay of temperature and pH changes, thus allowing the formulation of polymeric structures containing modified amines functionalities that exhibit dimensional response.[62]

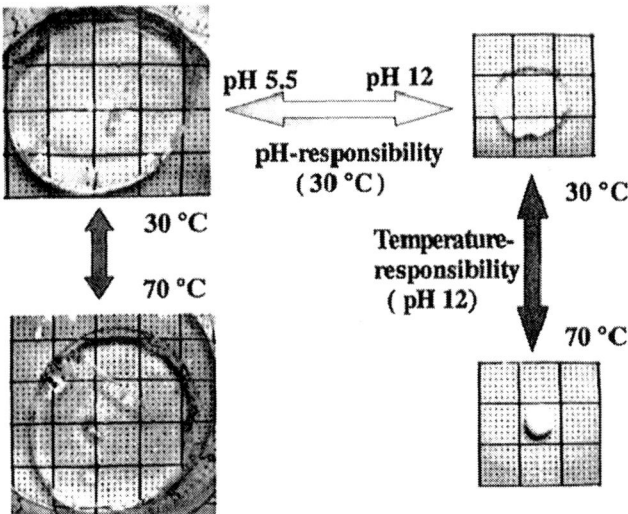

Figure 6. Swelling-shrinking behaviors of hydrolyzed copolymer hydrogels (Reproduced from reference 35. Copyright 2003 American Chemical Society.)

Polymer systems are relevant to physiological and biological systems due to their ability to respond to more than one stimulus, in particular pH and temperature.[63,64,65,66] Specifically, water-soluble stimuli-responsive polymers are of interest due to their applications in controlled drug delivery,[67] immobilized enzyme reactors,[68] and separation processes.[69] Stimuli such as pH, temperature, electric field, magnetic field, etc.,[70,71] cause the polymers, which are soluble in aqueous solutions, to undergo phase transitions. New types of thermal- and pH-responsive polymers have been synthesized from piperazine-based monomers and the increased chain length in the N-substituted alkyl group of the polymer increased its hydrophobicity.[72]

Changes in environment, pH,[73] ionic strength,[74] and temperature can cause certain chemically cross-linked polymer gels to undergo discontinuous volume-phase transitions and different degrees of swelling in solution will respond to different environmental changes. These types of materials may be used to design molecular sensing and actuation devices,[75] as well as other types of biomaterials and microelectronics. Through the restriction of the swelling transition of materials, stimuli-responsive surfaces and interfaces can be used to modulate cell attachment[76] and enable cell harvesting[77] and coculturing.[78] These effects have been studied on polyelectrolyte multilayers (PEMs),[79] which are commonly formed as dimensionally constrained thin films. The affinity of molecular species to PEMs can be spatially regulated by virtue of pH-triggered conformational changes in the PEM. These materials could provide a vehicle for studying the origin of volume phase transitions and multiple coexisting swollen phases in ultrathin films.[80]

Surface-Interfacial Phenomena

Exploitation of mixtures of materials may effectively and actively induce topography at the surface of soft polymeric films and coatings. Since in biological and biomedical materials the correlation between surface structures and properties is important, the same attributes hold for stimuli-responsive polymers. For example, surface structural features affect whether or not bacteria will grow on a particular surface or will be destroyed. When polymers are mixed with other materials, the size and the distribution of these entities will determine the overall responsive behaviors. For example, in nanocomposites and nanostructured materials, which often combine soft matter with metals, metal oxides, or ceramics for uses in electronics, packaging, or information storage, these properties are essential and also reflected in surface properties and will be reflected in surface topography. Although topography can be effectively examined using atomic force microscopy (AFM),[81] it is also equally important to determine chemical species responsible for surface morphological features which can be accomplished using recent advances in infrared imaging that overcame detection limits.[82]

Nanoporous membranes have been designed to mimic biological functions, and stimuli-responsive, "smart" materials have been designed through graft polymerization on these nanoporous membranes.[83] Responsiveness may result from pH or glucose concentration changes. For example, a pH-sensitive polypeptide brush was self-assembled on a nanoporous membrane and the nanometer-sized channels may open and close in response to pH changes. Figure 7 illustrates AFM images of pores that change their size as a function of pH. As shown, in the region of low

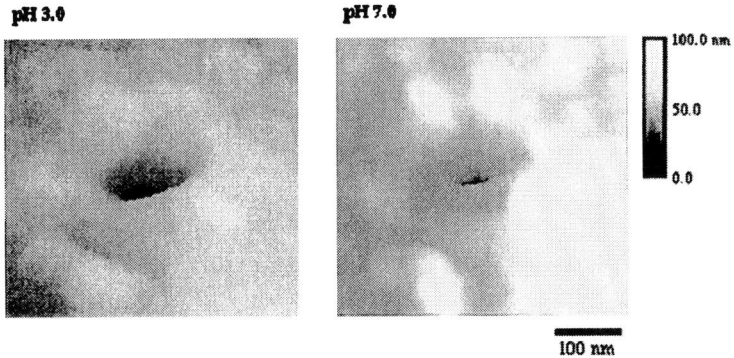

Figure 7. Self-assembled porous membranes at pH 3.0 and pH 7.0 (Reproduced from reference 99. Copyright 2000 American Chemical Society.) (See page 1 of color insert.)

pH, a poly(L-glutamic acid) (PLGA) chains are protonated and folded to form an α-helical structure. However, in a region exhibiting high pH, deprotonation occurs, leading to the formation of an extended random structure. Thus, these chemical changes result in permeation differences through the porous membrane which can be regulated by changing the pH due to direct contact of the polypeptide brush with the environment.

In order to prevent strong adhesion, release coatings are used at the interface between two materials and polyamines,[84] organic sulfides,[85] perfluorinated polyesters,[86,87] and poly(vinyl alcohol)s have been used to facilitate weak adhesion for pressure sensitive adhesives (PSA). However, for PSA release, silicone polymers are most frequently used due to low surface energy (20 mJ/m^2) and high flexibility.[88,89] By reacting an allyl amide functional perfluorinated ether (PFE) to a siloxane network through a hydrosilation reaction, the surface properties of a poly(dimethylsiloxane) (PDMS) elastomer can be effectively modified. The driving forces of lower energy groups segregating to the surface are capable of pulling high-energy groups to the free surface of a polymer, and when the surface reconstructs in contact with a high-energy environment, the high-energy groups may be used for further chemical interactions.[90]

Modification of synthetic surfaces with polymers[91,92,93] or self-assembled monolayers (SAMs),[94] is an effective method of controlling surface properties of interest. Surface-grafted stimuli-responsive, or "smart," polymers have shown the capability of these systems to reversibly switch wettability.[95] Upon synthesizing PNIPAm films on two-component alkanethiolate SAMs, the ratio of hydrophobic to hydrophilic thilates can be varied and the wettability and reversibility can be switched between a less hydrophobic state and a more hydrophobic state.[96] The ability to control interfacial free energy by being able to change it abruptly by application of an externally controlled stimulus may have useful applications in chromatographic separations,[97] cell culture,[98] and other forms of biotechnology.

When PLGA carrying a disulfide group at the amino terminal position (PLGA-SS) was prepared, it self-assembled on the surface of an Au coated nanoporous membranes and the pH dependence of water permeation varied with the length of the polypeptide chain, having lower sensitivity with longer PLGA-SS chains.[99]

The formation of multi stimuli-responsive membranes thus have a potential for numerous future applications an one of the examples is the development of a pulsatile local delivery system based on pH- and temperature-sensitive hydrogels that may be used to treat coronary thrombosis or stroke patients.[100]

SAMs or organic molecules containing thiol or disulfide moieties have useful applications in the areas of electronics,[101,102] catalysis[103,104], and biotechnology.[105,106,107] Previously, SAMs have been prepared through the absorption of low molecular weight molecules onto gold or other transition metals. Another route for the preparation of (co)polymer stabilized transition metal particles has been reported. RAFT is a versatile, controlled free radical polymerization process which allows for a wide range of (co)polymers with controlled molecular weights and architectures to become immobilized.[108]

Well-defined surfaces can be prepared through the deposition of SAMs[109] and the packing density is determined by the surface energy, stability, and resistance to surface reconstruction.[110,111] The combination of the self-assembly process with mechanical manipulation of grafted molecules on surfaces leads to the formation of mechanically assembled monolayers (MAMs) which are capable of forming superhydrophobic surfaces and exhibit long-lasting barrier properties. MAMs, fabricated with semi-fluorinated (SF) molecules have superior barrier properties and are considered superhydrophobic as well as nonpermeable. Subsequently, they do not deteriorate even after exposure to water.[112]

Ammonia/plasma treatments are capable of forming bonded NH functional groups on polymer surfaces. Nonbonded molecules in the substrate network may affect the gas plasma treatments as well. Plasma-surface modification has the ability to change a small surface without affecting the bulk properties of the polymer. PDMS elastomers, with and without chloro-functional species, have been ammonia/plasma-modified and showed that PDMS substrates which are free of chloro-functional molecules are capable of being amide-functionalized using ammonia/plasma modification. Apparently, the formation of an ammonium chloride surface layer inhibits surface-bonded amide groups from forming. A significant amount of surface ammonium chloride completely inhibits surface amide groups from forming as well as other surface functional groups.[113]

Synthetic polymers may be manipulated by changing the length, chemical composition, architecture, and topology of the surface chains, thus allowing the design of responsive materials triggered by external stimuli. In an effort to change the surface properties of PDMS, perfluoroalkanes can be attached to the surface of a cross-linked PDMS elastomer and the increase in the density of the perfluoroalkanes causes the contact angle with water to increase by ~30°, making the PDMS surface superhydrophobic.[114] Such diblock copolymers are composed of two chemically distinct polymers joined together at one end and microphase separation of the copolymer into ordered nanoscopic domains occurs due to the immiscibility of the two blocks.

Polymers are commonly used to produce high and low adhesion or friction in adhesive and lubricant coatings. Characterization of tribological properties has previously been limited to macroscopic or microscopic systems due to a lack of nanoscale techniques and the complexity of the systems themselves. Now, however, these properties can be understood at a molecular level with innovative computer devices and the miniaturization of machinery. For example, polystyrene (PS) and polyvinyl benzyl chloride (PVBC) can be used to examine the effect of polarity on friction and adhesion and for the two shearing polymer surfaces friction forces and polymer-polymer adhesion hysteresis are dependent on the rearrangement of the outermost polymer segments at shearing interfaces.[115]

Surface enrichment, surface-directed spinodal decomposition, critical point wetting, and adhesion are among those processes that are responsible for the behavior of simple fluid mixtures,[116, 117] complex fluids such as polymer blends and copolymers,[118,119,120,121,122,123,124] and numerous biological systems.[125] The behavior in these multi-component systems at surfaces and interfaces can differ significantly from that in the bulk due to one or more of the components having preferential affinities toward stratification near interfacial regions. Since interfacial energies of polymers at solid surfaces can be manipulated through controlling the chemical composition of the copolymer by synthesis and end-grafting statistical random copolymers onto the surface, surface properties can be also controlled. The wetting behavior of a homopolymer on a surface can be altered by varying the composition of a random copolymer grafted onto a surface thus allowing control of the interfacial behavior of simple fluids and to achieve biocompatibility through modification of the surface.[126]

Polymer surfaces may exhibit tacky/nontacky or hydrophilic-hydrophobic surfaces that can vary with surrounding conditions. Tackiness, or surface stickiness, and wettability, whether a liquid will spread on a surface as a continuous film or retract as one or several droplets, are both desirable properties in many applications. A material is considered sticky when the energy required to break its bond with a surface is significantly larger than the interfacial energy. Subsequently, a polymer is considered tacky when the desirable balance between softness and ability to dissipate energy is achieved. In contrast, wettability is more readily controlled by chemical nature and molecular organization of the surface because it is related to the surface interfacial energy. In order to control the wettability and tackiness of a material, various types of surface modifications were developed,[127, 128, 129, 130, 131] and one of the examples how tack and dewetting are affected by a smectic polymer liquid crystal structure have shown that the smectric structure brings hardness and nonwettability through the side-chain ordering. However, in the isotropic phase, tacky behavior and an ability to slow down the dynamics of wetting are seen due to the presence of the back-bone.[132]

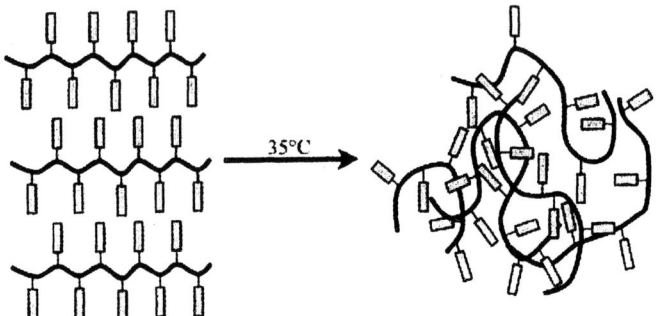

Figure 8. Schematic representation of the transition between the smectic and the isotropic phase as a result of temperature changes
(Reproduced with permission from reference 132. Copyright 1999 American Association for the Advancement of Science.)

Furthermore, as shown in Figure 8, one can control surface properties by temperature. In this case, at 35°C the network undergoes smectic-isotropic transition which results in wettability-tackiness changes. Thus, the bulk phenomenon is reflected in surface/interfacial property changes.

Nanostructures

Nanostructures can be generated through the synthesis of materials using nanostructured templates.[133] Subsequently, these templates consisting of microporous membranes,[134,135] zeolites,[136] or crystalline colloidal arrays.[137, 138,139] can be utilized to construct electronic, mechanical, or optical structures with stimuli-responsive characteristics. For example, porous Si is often used as a template[140] due to its ability of forming porous morphologies and, more importantly, the pore size to be tuned through adjusting the preparation conditions that allow the construction of optical structures.[141] Following the removal of the template through chemical dissolution, the polymer castings can be used as vapor sensors, optical fibers, and self-reporting bioresorbable materials.[142]

Formation of nano-structural features from inorganic silicon-containing triblock copolymers, displaying double gyroid and inverse double gyroid morphologies, were used to produce three-dimensional ceramic nanostructured films. Low-temperature technique using bifunctional oxidation processes, through careful selection of the relative volume fraction and phases, produced the porous and relief ceramic nanostructures from self-assembling, template-free block copolymer precursors. Through the selection of relative volume fraction and phases, block copolymers can be employed as precursors to prepare symmetries which allow nanostructures to be prepared.[143]

Metallocene-based polymer systems, such as polyferrocenylsilanes (PFSs), have been synthesized through ring-opening polymerization methods.[144,145] One example is pyrolysis which yields nanocomposites containing magnetic Fe clusters that retain the shape of the PFS precursor. In the pyrolysis process temperature controls and determines the size of the Fe nanoparticles, thus allowing the tuning in the magnetic properties from superparamagnetic to ferromagnetic.[146] These materials may be used for various applications such as data storage, precursors to electrostatically charged microspheres, detecting sensors in refractive indexes, and electromagnetic shielding applications.[147,148] Organic PS coblocks have also been prepared. However, PFS is more etch resistant than PS and therefore, thin films of these materials that lead to periodic nanoscale domains of PFS can be used for patterning substrates such as cobalt.[149]

Loading and distribution of nanoparticles in polymer matrixes play an important role to numerous bulk properties. However, it has been found that carbon particles may significantly reduce the adhesion between 2K polyurethanes (PUR) and acrylonitrile-butadiene-styrene (ABS) substrates. Apparently, an increase in the pigment volume content (PVC) reduces the adhesion to ABS due to stoichiometric imbalances resulting from the enhanced surface activity and surface area of carbon black particles which subsequently increase the inter-molecular H-bonding near the F-S interface in PUR coatings.[150]

Noncovalent interactions are used in the molecular self-assembly of supramolecular structures and materials. Hydrogen bonding and other weak reversible interactions are important for the design of new polymer architectures.[151,152] 2-ureido-4-pyrimidone, which dimerizes strongly in a self-complementary array of four cooperative hydrogen bonds, was used in reversible self-assembling polymer systems as the associating end group. This type of unidirectional design prevents multidirectional association or gelation. The polymer networks with thermodynamically controlled architectures can be developed for applications such as hot melts and coatings.[153]

Responses to Electric and Magnetic Fields

Although a number of external variables may affect special distribution of components in polymer networks, temperature and electric fields play a crucial role in the orientational ordering of molecules in the liquid crystalline (LC) state.[154,155] On the other hand, biological self-assembling systems are composed of a variety of discrete molecules forming heterogenous and hierarchical structures varying in length.

Hydrogen bonding and ionic interactions play a critical role in the formation of self-assembled liquid crystal (LC) structures, particularly the formation of supramolecular LC. Supramolecular materials are capable of responding to external stimuli by changing their self-assembled structures[156] and through the dissociation and association of noncovalent interactions. One example of self-organization and stimuli-responsive characteristics is a thermotropic folic acid derivative exhibiting change from a smectic to hexagonal columnar phases through the addition of alkali metal salts. As shown in Figure 9, the hydrogen-bonded pattern is changed from ribbon to disc due to ion-dipolar interactions that stabilize the structure. Changes in pH or ionic strength may also induce or destroy ordering of these structures.[104]

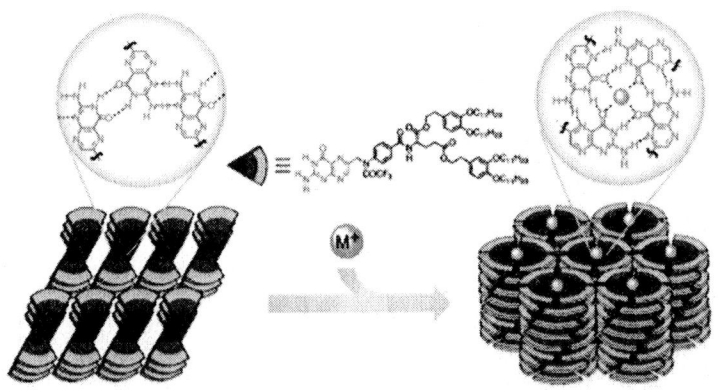

Figure 9. Ion-responsive self-assembled LC crystals with ribbon-like structures of a folic acid derivative and their structural change to the disc-like tetramers are obtained by the addition of metal ions (M^+) (Reproduced with permission from reference 167. Copyright 2002 American Association for the Advancement of Science.) (See page 1 of color insert.)

The use of self-assembling polymeric supramolecules is a powerful agent for producing materials that respond to external conditions. Combining recognition with self-organization allows for greater structural complexity and functionality. Hierarchically structured materials can be obtained through the application of self-organization and directed assembly. These types of materials are able to form a basis for tunable nanoporous materials, smart membranes, and anisotropic properties such as proton conductivity.[157]

A particular type of polymer brush, polyelectrolyte (PE) brushes, consists of electrically charged polymers.[158] Mixed PE brushes can be prepared by grafting using an esterification reaction and they are capable of switching the surface charge with a change in pH. When the top of the brush is occupied by negative charges stretched away from the substrate, or positively charged stretched poly(2-vinylpyridine) chains at pH values below the isoelectric point, the PE brush may exhibit switching behavior. This is schematically illustrated in Figure 10 and this type of responsive/switching behavior of the polyelectrolyte brush may result in tuning surface properties in an aqueous environment using a pH stimulus. Thus, there are numerous application opportunities for creating smart nanodevices, drug delivery systems, and microfluidics.[159]

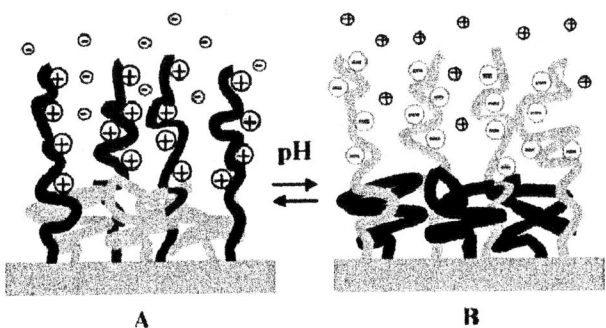

Figure 10. Schematic representation of the switching behavior of mixed polyelectrolyte brushes upon change of pH below the isoelectric point (A) and above the isoelectric point (B) (Reproduced from reference 159. Copyright 2003 American Chemical Society.)

Polymer brushes are polymer chains which are tethered to a surface or an interface with such high grafting density that the chains are forced to stretch away from the tethering site.[160] These tethered chains can be formed by polymer physisorption, where a diblock copolymer is used to strongly absorb to the surface and stretch away to form the brush layer.[161] The chains can also be formed by the chemical bonding of chains to the interface. Through the combination of SAM, formation and living free radical polymerization techniques, the thickness and architecture of the polymer brush layer can be controlled. Semifluorinated diblock copolymer brushes from silica substrates were synthesized using surface-initiated atom transfer radical polymerization (ATRP) techniques.[162] These types of fluoropolymer materials are biocompatible, and can be used for medical applications and drug delivery.[163]

LC materials are liquids that possess anisotropic optical and electrical properties, and arise from the preferable orientation of molecules within the liquid, and may spontaneously assume orientations that depend on topography and chemical functionality of the surfaces. Generally, the application of an electric field across LCs orientated by a surface will change the orientation of most of the LC, but not the regions near the surface.[164,165] However, a method to electrically drive the orientation of the LCs from the surface through electrochemical control of the oxidation state of ferrocene-decorated electrodes has been shown to cause surface-driven

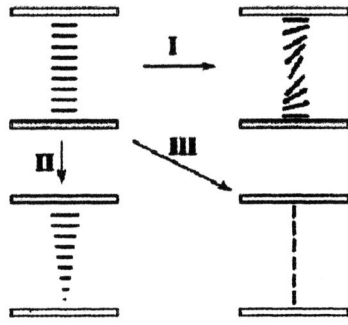

Figure 11. Schematic illustration of changes in orientations of LCs: (I) Change driven by the action of an electric field on the bulk of the LC; (II and III) Surface-driven changes in orientation leading to either formation of a helical twist in the LC (II) or an orientation of the liquid crystal that is perpendicular to the surfaces (III) (Reproduced with permission from reference 166. Copyright 2003 American Association for the Advancement of Science.)

changes in the LCs orientations.[166] As illustrated in Figure 11, the orientation of LCs induced by electric fields changes from parallel to perpendicular to the surface.

The applicability of LCs in high-tech and bio related fields may be widened through the creation of complex phase-segregated structures. The creation of more complex hierarchical structures with liquid crystals yields functionality on a molecular level, such as transportation and sensing of electrons, ions, and molecules.[167]

Another interesting example is azobenzene which is a photoresponsive chromophore, with isomerizations that are photo- and thermal-induced. These species and its derivatives are able to take *trans* and *cis* structures with respect to the azo-linkage, but due to stability, typically exist in the trans form. Poly(phenylene)-based conjugated polymers with photoisomerizable azobenzene units in the main chain using palladium-catalyzed coupling have been prepared and their electrochemical properties change due to the isomerization. Their stimuli-responsive nature results from the built-in conjugated polymer backbone that may serve as molecular sensors and optical switching devices.[168] It has been proposed that smart materials may be designed through the photo- and
thermo-regulated conformational changes in azobenzene modified polymers and in order to construct high-performance materials with stimuli-responsive chiro-optical behavior, a series of azobenzene modified polymers

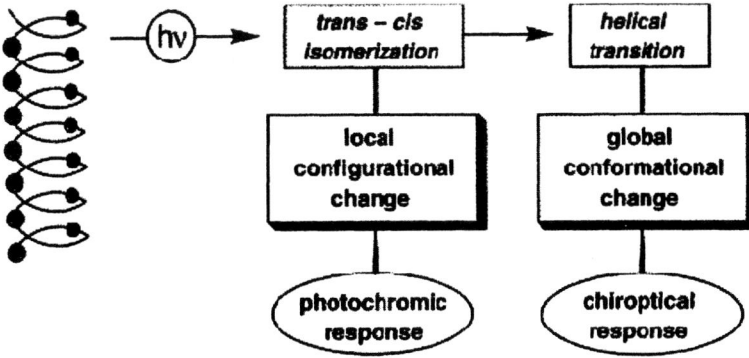

Figure 12. Coupled photochromic and chiroptical responses in azobenzene modified helical polymers (Reproduced with permission from reference 169. Copyright 2001 American Chemical Society.)

and their derivative model compounds were fitted with axially asymmetric R- or S-binaphthyl groups. Subsequently, near-ultraviolet light and heat is capable of tuning the chiro-optical properties which is manifested by the diminished circular dichroism and optical rotation intensities. It turns out that these effects are fully reversible, and are attributed to the helical conformations in the trans-azobenzene modified polymers that were disrupted following the *trans-cis* isomerization reaction.[169] These cumulative effects of *trans-cis* isomerization and helic transitions are illustrated in Figure 12.

In 1968, Mataga proposed an alternative approach to organic magnets which is based on π-conjugated polymers.[170] This attractive macromolecular approach may be used to obtain magnetic properties at room temperature as well as at high temperatures and is attributed to the fact that the exchange interactions between electron spin through the π-conjugated systems can be made stronger than the through-space interactions in molecular solids of organic radicals.[171] To achieve these properties it is necessary to design a polymer with a large magnetic moment and magnetic order at low temperatures. With a large crosslink density and alternating connectivity of radical modules with unequal spin quantum numbers, such polymer acts as an effective magnetic moment and slow reorientation of magnetization by a small magnetic field below a temperature of about 10 K.[172]

High-performance ceramics such as SiC, AlN, B_4N, and TiN have many practical applications in disciplines ranging from rocket nozzles to automotive parts.[173, 174, 175] Particularly, ceramics obtained as film, coatings, bulk shapes, or fibers are especially useful. Recent studies suggest that organosilicon polymers and inorganic glasses may serve as good precursors for ceramic materials, as they have a three-dimensional network of covalent bonds and can be easily processed.[176,177,178,179,180] Synthesis by thermal polymerization has yielded a metal-containing polymer from which a shaped, magnetic ceramic was obtained and by controlling the pyrolysis conditions, the magnetic properties of the shaped ceramic could be tuned between a superparamagnetic and ferromagnetic state. Therefore, the cross-linked metal-containing polymers may serve as useful precursors to ceramic monoliths with tailorable magnetic properties which can be molded and may possess magnetic, electrical, and optical properties.[181] The metal-containing organosilicon polymer networks may function in iron delivery as well as serving as a matrix for the nucleation of iron nanoclusters.

In summary, this chapter introduced only a few essential aspects of the ongoing efforts in the growing field of stimuli-responsive polymeric films and coatings. There are many emerging areas including photonics, nano-composites, biological and biomedical areas, just to name a few, that will directly or indirectly benefit from the ongoing efforts in the development of new and adopting existing materials with stimuli-responsive characteristics. The following chapters focus on specific synthetic and chemico- physical aspects in further advancing our knowledge in this important area of research.

Acknowledgments

This work was partially supported by the National Science Foundation Materials Research Science and Engineering Program (DMR 0213883).

References

1. Morice, M. *N.Engl. J. Med.*, **2002**, 346, 1773.
2. Langer, R. Nature, **1998**, 392, 5.
3. Yasuda, H., Stannet, V., Frisch, H.L., Peterlin, A. *Makromol. Chem.* **1964**, 73, 188.
4. Heffelfinger, C.J. *Polym. Eng. Sci.* **1978**, 18, 1163.
5. Stach, W., Holland-Moritz, K. *J. Mol. Struc.* **1980**, 60, 49.
6. Siesler, H.W. Markomol. Chem., **1979**, 180, 2261.
7. Urban, M.W., Chatzi, Perry, B.C., Koenig, J.L. Appl. Spectrosc., **1986**, 40, 1103.
8. Caddock, B.D., Evans, K.E. *J. Phys. D: Appl. Phys.* **1989**, 22, 1877.
9. Evans, K.E., Caddock, B.D. *J. Phys. D: Appl. Phys.* **1989**, 22, 1883.
10. Ludwig, B.W., Urban, M.W. *Polymer* **1994**, 35, 5130.
11. Ludwig, B.W., Urban, M.W. *Polymer* **1992**, 33, 3343.
12. Lendlein, A.; Langer, R, *Science*, **2002**, 296, 1673.
13. Hunter, J.G, Ed. *Minimally Invasive Surgery* **1993**.
14. Lendlein, A., Langer, R. Science **2002**, 296, 1673-1676.
15. Charlesby, A. *Atomic Radiation and Polymers* **1960**, 198-257.
16. Kagami, Y., Gong, J.P., Osada, Y. *Macromol. Rapid Commun.* **1996**, 17, 539.

17. Kim, B.K., Lee, S.Y., Xu, M. *Polymer* **1996**, 37, 5781.
18. Lin, J.R., Chen, W. *J. Appl. Polym. Sci.* **1998**, 69, 1563.
19. Takahashi, T., Hayashi, N., Hayashi, S. *J. Appl. Polym. Sci.* **1996**, 60, 1061.
20. Sakurai, K., Shirakawa, Y., Kahiwagi, T., Takahashi, T. *Polymer* **1994**, 35, 4238.
21. Sakurai, K., Takahashi, T. *J. Appl. Polym. Sci.* **1989**, 38, 1191.
22. Sakurai, K., Takahashi, T. *J. Appl. Polym. Sci.* **1993**, 47, 937.
23. Osada, Y., Matsuda, A. *Nature* **1995**, 376, 219.
24. Hu, Z., Zhang, X., Li, Y. *Science* **1995**, 269, 525.
25. Schetky, L.M. *Sci. Am.* **1979**, 241, 68.
26. Swain, M.V. *Nature* **1986**, 322, 234.
27. Goodman, S.H., Ed. *Handbook of Thermoset Plastics* **1998**.
28. Kaiser, T. *Prog. Polym. Sci.* **1989**, 14, 373.
29. Sauer, J.A., Hara, M. *Adv. Polym. Sci.* **1990**, 91/92, 69.
30. White, S.R., et al. *Nature* **2001**, 409, 794.
31. Talreja, R., Ed. *Damage Mechanics of Composite Materials* **1994**, 139.
32. Kinloch, A.J. *Adv. Polym. Sci.* **1985**, 72, 45.
33. Chen, X., et al. *Science* **2002**, 295, 1698-1702.
34. Glass, E.J. Ed. Symp. Series #213, *American Chemcial Society*, 1986.
35. Babcock, H.P., Teixeira, R.E., Hur, J.S., Shaqfeh, E.S.G., Chu, S. *Macromolecules* **2003**, 36, 4544.
36. Schroeder, C.M, Babcock, H.P., Shaqfeh, E.S.G., Chu, S. *Science* **2003**, 301, 1515-1519.
37. Benee, L.S.; Snowden, M.J.; Chowdhry, B.Z. *Langmuir*, **2002**, 18, 6025.
38. Kawaguchi, H., Fujimoto, K., Mizuhara, Y. *Colloid Polym. Sci.* **1992**, 270, 53.
39. (a) Snowden, M. J.; Thomas, D.; Vincent, B. *Analyst* **1993**, 118, 1367. (b) Morris, G. E.; Vincent, B.; Snowden, M. J. *J. Colloid Interface Sci.* **1997**, 190, 198.
40. Kiminta, D. M. O.; Luckham, P. F. *Polymer* **1995**, 36, 4827.
41. Rodriguez, B. E., Wolfe, M. S.; Fryd, M. *Macromolecules* **1994**, 27, 6642.
42. Saunders, B. R.; Crowther, H. M.; Vincent, B. *Macromolecules* **1997**, 30, 482.
43. Ma, G. H.; Fukutomi, T. *Macromolecules* **1992**, 25, 1870.
44. Ma, G. H.; Fukutomi, T.; Nozaki, S. *J. Appl. Polym. Sci.* **1993**, 47, 1243.
45. Loxley, A., Vincent, B. *Colloid Polym. Sci.* **1997**, 275, 1108.
46. Kuckling, D. Vo, C. D., Wohlrab, S. E. *Langmuir* **2002**, *18*, 4263.
47. Murthy, N.; Xu, M.; Schuck, S.; Kunisawa, J.; Shastri, N.; Frechet, J. M. J. *Proc. Natl. Acad. Sci.* **2003**, *100*, 4995.
48. Amalvy, J. I., Wanless, E. J., Li, Y., Michailidou, V., Armes, S.P. *Langmuir*, **2004**, 20, 8992.
49. Schilli, C.M., Zhang, M., Rizzerdo, E., Thang, S.H., Chong, Y.K., Edwards, K., Karlsson, G., Muller, A.H.E., *Macromolecules*, **2004**, 37, 7861.

50. Wichterle, O.; Lim, D. *Nature (London)* **1960**, *185*, 118.
51. Barrett, G. D.; Constable, I. J.; Steward, A. D. *J. Cataract* Refract. Surg. **1986**, 12, 23.
52. Menace, R.; Skorpik, C.; Juchem, M.; Scheidel, W.; Schranz, R. J. J. Cataract Refract. Surg. **1989**, 15, 264.
53. Seifert, L. M.; Green, R. T. *J. Biomed. Mater. Res.* **1985**, *19*, 1043.
54. Okano, T.; Aoyagi, T.; Kataoka, K.; Abe, K.; Sakurai, Y.; Shimadada, M.; Shinohara, I. *J. Biomed. Mater. Res.* **1986**, *20*, 919.
55. Weaver, J.V.M., Bannister, I., Robinson, K.L., Bories-Azeau, X., Armes, S.P., Smallridge, M., McKenna, P., *Macromolecules* **2004**, 37, 2395.
56. Kobayashi, S., Suh, K.D., Shirokura, Y. *Macromolecules*, **1989**, 22, 2363.
57. Tbal, H. Maguer, D.L., Morcellet, J., Delporte, M., Morcellet, M. *React. Polym.* **1992**, 17, 207.
58. Moriuchi, F., Muroi, H., *Jpn. Kokai Tokkyo Koho* **1987**, JP 62011093.
59. Anderson, N.J., Blesing, N.V., Bolto, B.A., Jackson, M.B. *React. Polym.* **1987**, 7, 17.
60. Qiu, Y., Zhang, T., Ruegsegger, M., Marchant, R.E. *Macromolecules* **1998**, 31, 165.
61. Martel, B., Pollet, A., Morcellet, M. *Macromolecules* **1994**, 27, 5258.
62. Yamamoto, K., Serizawa, T., Muraoka, Y., Akashi, M. *Macromolecules* **2001**, 34, 8014-8020.
63. Chen, G., Hoffman, A.S. *Nature* **1995**, 373, 49.
64. Feil, H., Bae, Y.H., Feijen, J., Kim, S.W. *Macromolecules* **1992**, 25, 5528.
65. Cho, S.H., Jhon, M.S., Yuk, S.H., Lee, H.B. *J. Polym. Sci. Part B: Polym. Phys.* **1997**, 35, 595.
66. Yuk, S.H., Cho, S.H., Lee, S.H. *Macromolecules* **1997**, 30, 6856.
67. Hoffman, A.S. In *Controlled Drug Delivery: Challenges and Strategies* **1997**, 485.
68. Galaev, Y.I. *Russ. Chem. Rev.* **1995**, 64, 471.
69. Park, T.G., Hoffman, A.S. *Appl. Biochem. Biotechnol.* **1988**, 19, 1.
70. Okahata, Y., Noguchi, H., Seki, T. *Macromolecules* **1986**, 19, 493.
71. Frank, S., Lauterbur, P. *Nature* **1993**, 363, 334.
72. Gan, L.H., Gan, Y.Y., Deen, R.G. *Macromolecules* **2000**, 33, 7893.
73. Annaka, M., Tanaka, T. *Nature (London)* **1992**, 355, 430-432.
74. Ohmine, I., Tanaka, T. *J. Chem. Phys.* **1982**, 77, 5725-5729.
75. Rogers, C.A. *Sci. Am.* **1995**, 273, 154-157.
76. Ito, Y., Chen, G.P., Guan, Y.Q., Imanishi, Y. *Langmuir* **1997**, 13, 2756.
77. Okano, T., Yamada, N., Sakai, H., Sakurai, Y. *J. Biomed. Mater. Res.* **1993**, 27, 1243-1251.
78. Hirose, M., et al. *Yonsei Med. J.* **2000**, 41, 803-813.

79. Decher, G. *Science* **1997**, 277, 1232-1237.
80. Hiller, J., Rubner, M.F. *Macromolecules* **2003**, 36, 4078-4083.
81. Assender, H., Bliznyuk, V., Porfyrakis, K. *Science* **2002**, 297, 973-976.
82. Otts, D., Zhang, P., Urban, M.Q., Langmuir, 2002,XXX
83. Osada, Y., Honda, K., Ohta, M. *J. Membr. Sci.* **1986**, 27, 327.
84. Tsujinaka, K., Akasaki, I. US Patent 5,633,328, **1997**.
85. Scttincri, R., O'Brien, J. US Patent 3,928,690, **1975**.
86. Dabroski, W. US Patent 4,513,059, **1985**.
87. Tynan, J. US Patent 5,451,440, **1995**.
88. Owen, M.J. *J Coatings Technol.* **1981**, 53, 49.
89. Tobolsky, A.V. In *Properties and Structures of Polymers*, **1960**, 67.
90. Thanawala, S.K., Chaudhury, M.K. *Langmuir* **2000**, 12, 1256-1260.
91. Zhu, X.Y., et al. *Langmuir* **2001**, 17, 7798
92. Ista, L.K., Perez-Luna, V.H., Lopez, G.P. *Appl. Environ. Microbiol.* **1999**, 65, 1603.
93. Leckband, D., Sheth, S., Halperin, A. *J. Biomater. Sci., Polym. Ed.* **1999**, 10, 1125.
94. Bain, C.D., Evall, J., Whitesides, G.M. *J. Am. Chem. Soc.* **1989**, 111, 7155.
95. Takei, Y.G., Aoki, T., Sanui, K., Ogata, N., Sakurai, Y., Okano, T. *Macromolecules* **1994**, 27, 6163.
96. Mendez, S., Ista, L.K., Lopez, G.P. *Langmuir* **2003**, 19, 8115-8116.
97. Kanazawa, H., Sunamoto, T., Ayano, E., Matsushima, Y., Kikuchi, A., Okano, T. *Anal. Sci.* **2002**, 18, 45.
98. Okano, T., Yamada, N., Okuhara, M., Sakai, H., Sakurai, Y. *Biomaterials* **1995**, 16, 297.
99. Ito, Y., Park, Y.S., Imanishi, Y. *Langmuir* **2000**, 16, 5376-5381.
100. Brazel, C.S., Peppas, N.A. *J. Controlled Release* **1996**, 39, 57.
101. Osifchin, R.G., Andres, R.P., Henderson, J.I., Kubiak, C.P., Dominey, R.N. *Nanotechnology* **1996**, 7, 412.
102. Sato, T., Ahmed, H., Brown, D., Johnson, B.F.G. *Appl. Phys.* **1997**, 82, 696.
103. Zhao, M., Crooks, R.M. *Angew.Chem., Inter. Ed.* **1999**, 38, 364.
104. Zhao, M., Crooks, R.M. *Adv. Mater.* **1999**, 11, 217.
105. De la Fuente, J., et al. *Angew. Chem., Int. Ed.* **2001**, 40, 2257.
106. Coa, Y.W., Jin, R., Mirkin, C.A., *J. Am. Chem. Soc.* **2001**, 123, 7961.
107. Li, Z., Jin, R., Mirkin, C.A., Letsinger, R.L. *Nucleic Acids Res.* **2002**, 30, 1558.
108. Sumerlin, B., Lowe, A., Stroud, P., Zhang, P., Urban, M.W. McCormick, C.L. *Langmuir* **2003**, 19, 5559-5562.
109. Chaudhury, M.K. *Mat. Sci. Eng. Rep.* **1996**, 16, 97.
110. Ulman, A. *An Introduction to Ultrathin Organic Films from Langmuir-Blodgett to Self Assembly* **1991**.
111. Wang, J., Mao, G., Ober, C.K., Kramer, E.J. *Macromolecules* **1997**, 30, 1906.

112. Genzer, J., Efimenko, K. *Science* **2000**, 290, 2130-2133.
113. Gaboury, S.R., Urban, M.W. *J. Appl. Poly. Sci.* **1992**, 44, 401-407.
114. Genzer, J., Efimenko, K. *Science* **2000**, 290, 2130.
115. Maeda, N., Chen, N., Tirrell, M., Israelachvili, J.N. *Science* **2002**, 297, 379.
116. Cahn, J.W. *J. Chem. Phys.* **1997**, 66, 3667.
117. Moldover, M.R., Cahn, J.W. *Science* **1980**, 207, 1073.
118. Wiltzius, P., Cumming, A. *Phys. Rev. Lett.* **1991**, 66, 3000.
119. Jones, R.A.L., et al. *ibid*, 1326.
120. Zhao, W., et al. *Macromolecules* **1991**, 24, 5991.
121. Norton, L.J., et al. *ibid* **1995**, 28, 8621.
122. Henkee, C.S., Thomas, E.L., Fetter, L.J. *J. Mat. Sci.* **1988**, 23, 1685.
123. Coulon, G., Deline, V.R., Green, P.F., Russell, T.P. *Macromolecules* **1989**, 22, 2581.
124. Anastasiadis, S.H., Russell, T.P., Satja, S.K., Majkrzak, C.F. *Phys Rev. Lett.* **1989**, 62, 1852.
125. Nojiri, C., et al. *J. Biomed. Res.* **1990**, 24, 1151.
126. Manksy, P., Liu, Y., Huang, E., Russell, T.P., Hawker, C. *Science* **1997**, 275, 1458-1460.
127. Langmuir, I. *Trans. Faraday Soc.* **1920**, 15, 62.
128. Pittman, A.G. *Fluoropolymers* **1972**, 25.
129. Maoz, R., Sagiv, J. *J. Colloid Interface Sci.* **1984**, 100, 465.
130. Bain, C.D., Evall, J., Whitesides, G.M. *J. Am. Chem. Soc.* **1989**, 111, 7155.
131. Mansky, P., Liu, Y., Huang, E., Russell, T.P., Hawker, C. *Science* **1997**, 275, 1458.
132. de Crevoisier, G., Fabre, P., Corpart, J., Leibler, L. Science 1999, 285, 1246-1249.
133. Polarz, S., Antonietti, M. *J. Chem. Soc. Chem. Commun.* **2002**, 2002, 2593.
134. Wirtz, M., Parker, M., Kobayashi, Y., Martin, C.R. *Chem. Eur. J.* **2002**, 16, 3572.
135. Hulteen, J.C., Martin, C.R. *J. Mater. Chem.* **1997**, 7, 1075.
136. Moller, K., Bein, T. *Chem. Mater.* **1998**, 10, 2950.
137. Reese, C.E., Baltusavich, M.E., Keim, J.P., Asher, S.A. *Anal. Chem.* **2001**, 73, 5038.
138. Xu, X., Majetich, S.A., Asher, S.A. *J. Am. Chem. Soc.* **2002**, 124, 13864.
139. Haynes, C.L, Van Duyne, R.P. *J. Phys. Chem.* **2001**, 105, 5599.
140. Matthias, S. *Adv. Mater.* **2002**, 14, 1618.
141. Thonissen, M., Berger, M.G. *Properties of Porous Silicon* **1997**, 18, 30.
142. Li, Y.Y., et al. *Science* **2003**, 299, 2045-2046.
143. Chan, V., et al. *Science* **1999**, 286, 1716.
144. Kingsborough, R.P., Swager, T.M. *Prog. Inorg. Chem.* **1999**, 48, 123.
145. Foucher, D.A., Tang, B.Z., Manners, I. *J. Am. Chem. Soc.* **1992**, 114, 6246.

146. MacLachlan, M.J., et al. *Science* **2000**, 287, 1460.
147. Kulbab, K., Manners, I. *Macromol. Rapid. Commun.* **2001**, 22, 711.
148. Espada, L. I., et al. *Inorg. Organomet. Polym.* **2000**, 10, 169.
149. Manners, I. *Science* **2001**, 294, 1665-1666.
150. Kalpana, K.S., Urban, M.W. *J. Coat. Techn.* **2000**, 72, 903.
151. Hilger, C., Drager, M., Stadler, R. *Macromolecules* **1992**, 26, 2498.
152. Aggeli, A. *Nature* **1997**, 386, 259.
153. Sijbesma, R.P., et al. *Science* **1997**, 278, 1601-1604.
154. Demus, D., Goodby, J.W., Gray, G.W., Spiess, H.-W., Vills, V., Eds. *Handbook of Liquid Crystals* **1998**.
155. Goodby, J.W. *Curr. Opin. Solid State Mater. Sci* **1999**, 4, 361.
156. Kanie, K., et al. J. Mater. Chem. **2001**, 11, 2875.
157. Ikkala, O., ten Brinke, G. *Science* **2002**, 295, 2407-2409.
158. Biesalki, M., Ruhe, J., Kugler, R., Knoll, W. In *Handbook of Polyelectrolytes* **2002**.
159. Houbenov, N., Minko, S., Stamm, M. *Macromolecules* **2003**, 36, 5897.
160. Milner, S.T. *Science* **1991**, 251, 905.
161. Belder, G.F., ten Brinke, G., Hadziioannou, G. *Langmuir* **1997**, 13, 4102.
162. Granville, A.M, Boyes, S.G., Akgun, B., Foster, M.D., Brittain, W.J. *Macromolecules* **2004**, 37, 2790-2796.
163. Zhang, Z., Yang, S., Zhang, Q., Xu, X. *J. Polym. Sci., Part A: Polym. Chem.* **2001**, 39, 2670.
164. Collings, P.J., Patel, J.S. *Handbook of Liquid Crystal Research* **1997**.
165. Jerome, B. Rep. *Prog. Phys.* **1991**, 54, 391.
166. Luk, Y., Abott, N.L. *Science* **2003**, 301, 623-626.
167. Kato, T. *Science* **2002**, 295, 2414- 2418.
168. Izumi, A., Nomura, R., Masuda, T. *Macromolecules* **2001**, 34, 4342.
169. Lustig, S.R., Everlof, G.J., Jaycox, G.D. *Macromolecules* **2001**, 2364.
170. Mataga, N. *Thero. Chim. Acta* **1968**, 10, 372.
171. Rajca, A. *Chem. Rev.* **1994**, 10, 871.
172. Rajca, A., Wongsriratanakul, J., Rajca, S. *Science* **2001**, 294, 1503.
173. Baldus, H.-P., Jansen, M. *Angew Chem. Int. Ed. Engl.* **1997**, 36, 328.
174. Bill, J., Aldinger, F. *Adv. Mater.* **1995**, 7, 775.
175. Segal, D. *Chemical Synthesis of Advanced Ceramic Material* **1991**.
176. Peuckert, M., Vaahs, T., Bruck, M. *Adv. Mater.* **1990**, 2, 398.
177. Laine, R.M., Babonneau, F. Chem Mater. 1993, 5, 260.
178. Wideman, T., Remsen, E.E., Cortez, E., Chlanda, V.L. Sneddon, L.G. *Chem Mater.* **1998**, 10,412.
179. Visscher, G.T., Nesting, D.C., Badding, J.V., Bianconi, P.A. *Science* **1993**, 260, 1496.
180. He, J., Scarlete, M., Harrod, J.F. *J. Am. Ceram. Soc.* **1995**, 78, 3009.
181. MacLachlan, M.J. *Science* **2000**, 287, 1460-1463.

Synthelic Aspects of Stimuli-Responsive Polymers

Chapter 2

Controlling Polymer Chain Topology and Architecture by ATRP from Flat Surfaces

Joanna Pietrasik, Lindsay Bombalski, Brian Cusick, Jinyu Huang, Jeffrey Pyun, Tomasz Kowalewski, and Krzysztof Matyjaszewski

Department of Chemistry, Carnegie Mellon University, 4400 Fifth Avenue, Pittsburgh, PA 15213

Polymer brushes prepared by end grafting of chains onto/from flat surfaces were synthesized by atom transfer radical polymerization. This process enables the preparation of wide range of brushes with precise molar mass, composition and architecture. By tuning parameters such as grafting density, chemical composition or type of applied surfaces, products with tailored properties could be achieved. Several approaches to synthesize stimuli-responsive polymeric brushes by ATRP are presented.

Modification of surfaces with thin polymer films is commonly used to tailor surface properties such as wettability, biocompatibility, adhesion, adsorption, corrosion resistance and friction. Such thin polymer films can be applied by depositing or spraying a polymeric coating from solution. Alternatively, polymers with reactive end groups can be grafted onto surfaces, resulting in polymer brushes.[1,2]

Polymer brushes are defined as dense layers of chains confined to a surface or interface where the distance between grafts is much less than the unperturbed dimensions of the tethered polymer.[3,4] Due to high steric crowding, grafted chains extend from the surface resulting in an entropically unfavorable conformation. Polymer brushes have been prepared by the end-grafting of chains onto/from flat or curved surfaces that are organics, or inorganic in nature. These include functional colloids, highly branched polymers and block copolymer aggregates, such as micelles or phase separated nanostructures.[3,4] Because of kinetic and thermodynamic restrictions, grafting-onto approach lead to formation of polymer brushes with relatively low grafting density.[5]

Atom transfer radical polymerization (ATRP) has been demonstrated to be a versatile technique to synthesize well-defined polymers, as well as complex (co)polymers and organic/inorganic hybrid materials because of ATRP compatibility with a wide range of functional monomers.[6-12] Variation of polymer brush composition and degree of polymerization (DP) using ATRP has been extensively investigated to modify surface properties, nanopattern and design stimuli-responsive materials.[13-19] Stimuli responsive materials undergo a reversible phase transition which causes changes in polymer conformation and polymer surface energetics. External stimuli, such as pH, nature of solvent, temperature, light or an electric field, allow for fine-tuning of the surface properties of specially designed responsive polymer materials.[20,21]

ATRP of various monomers from preformed polymeric or colloidal initiators has been achieved yielding nano-objects of precise dimensions and functionality.[22-26] Fundamental investigations into the parameters affecting surface initiated ATRP have been focused on exploring the conditions for controlling film thickness, functionality and properties. Recent work also demonstrated the ability to control grafted polymer architecture ranging from tethered linear and hyperbranched polymers to network crosslinked films.[27,28] In the characterization of these systems, techniques, such as, ellipsometry, contact angle, XPS and AFM are central in assessing whether tethered (co)polymers obtained from surface initiated ATRP possess precise molar mass and composition.

This article focuses on stimuli responsive polymer brushes synthesized in our laboratory using ATRP. Discussed herein polymer brushes were attached to the surfaces by using either grafting-onto or grafting-from approach.

Stimuli-Responsive Ultrathin Films by "Grafting-onto" Approach

In order to exhibit distinct morphological differences due to stimulus-driven rearrangement by treatment with selective stimuli i.e., solvents, pH, temperature, salts, in addition to appropriate compositions, copolymer brushes must possess sufficient conformation freedom. This prompted to design a stimulus-responsive brush utilizing a "grafting-onto" approach of a functional ABC triblock copolymer. Due to the low grafting densities inherent to this synthetic method, tethered copolymers retained adequate degrees of freedom enabling segment selective behavior to various solvents.[29]

The synthesis of reactive ABC triblock copolymer precursor was conducted by a combination of living anionic ring-opening and atom transfer radical polymerizations.[2,30] In the first step, a poly(dimethylsiloxane) pDMS macroinitiator ($M_{n\ SEC}$ = 6 200; M_w/M_n = 1.19) was prepared by the living anionic ring-opening polymerization and hydrosilation of the silane end-group with an allyl 2-bromoisobutyrate. Chain extension from the pDMS macroinitiator using ATRP of styrene (St) yielded a pDMS-b-pSt diblock copolymer (M_n = 66 730; M_w/M_n = 1.38). A final ATRP step, using the diblock copolymer macroinitiator with 1-(dimethoxymethylsilyl)propyl acrylate (DMSA), yielded a triblock copolymer ($M_{n\ NMR\ pDMS-b-pSt-b-pDMSA}$ = 156 700) capable of covalent bonding to a silicon wafer due to presence of alkoxy groups of copolymer chains, also containing rubbery and glassy segments. Treatment with toluene, or a mixture of toluene/hexane allowed for the reversible control of surface properties of the copolymer ultra-thin films, since it led to the surface presentation of either pSt or pDMS segments. Surface morphology of films which were immersed in a good solvent for both segments (toluene) and subsequently dried under a stream of nitrogen was similar to a typical fractal morphology of glassy polymers, indicating that pSt segments were predominantly presented at the surface (Figure 1 *left*). In contrast, as shown in Figure 1 *right* top, exposure to toluene-hexane mixtures and drying under nitrogen, rendered surfaces which were completely featureless when imaged under ultra-light tapping conditions ($A/A_o \rightarrow 1$, where A and A_o denote, respectively, the set-point and "free" cantilever oscillation amplitude). Such behavior indicates that "soft" pDMS segments preferentially segregated to the surface following the treatment with the solvent of lower affinity towards pSt. Imaging under "normal" tapping conditions (A/A_o=0.9), revealed the globular morphology, pointing to the presence of collapsed glassy pSt domains under the PDMS layer (Figure 1 *right*, bottom).

Similar approach was applied for the synthesis of diblock copolymer containing poly(N,N-dimethylaminoethyl methacrylate) (M_n = 14 600, M_w/M_n = 1.07) which was extended with poly(trimethoxysilylpropyl methacrylate) block poly(DMAEMA$_{93}$-b-TMSPMA$_{36}$), (Molecular weight (M_n) determined by comparing the peak areas of the protons derived from DMAEMA and TMSPMA segments in ^{1}H NMR spectrum M_n = 24 000 g/mol) (Scheme 1).

Surface properties of the copolymer ultrathin films could be reversibly controlled by treatment with selective solvents or temperature as pDMAEMA exhibits low critical solution temperature (LCST). PDMAEMA chains of

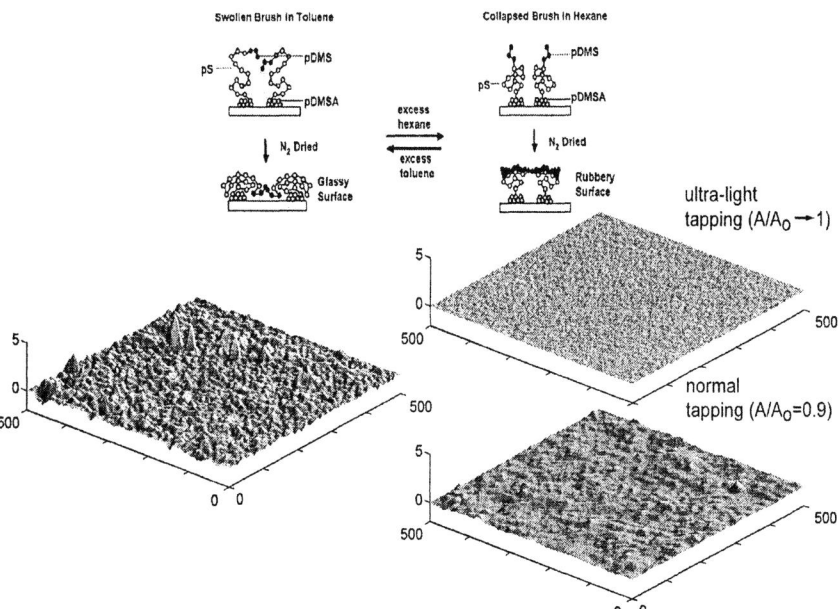

Figure 1: Tapping mode AFM height images of pDMS-b-pSt-b-pDMSA brushes after different solvent treatments followed by drying with nitrogen. Left after immersion in toluene; right after immersion in toluene and gradual addition of hexane. (All scales in nm).

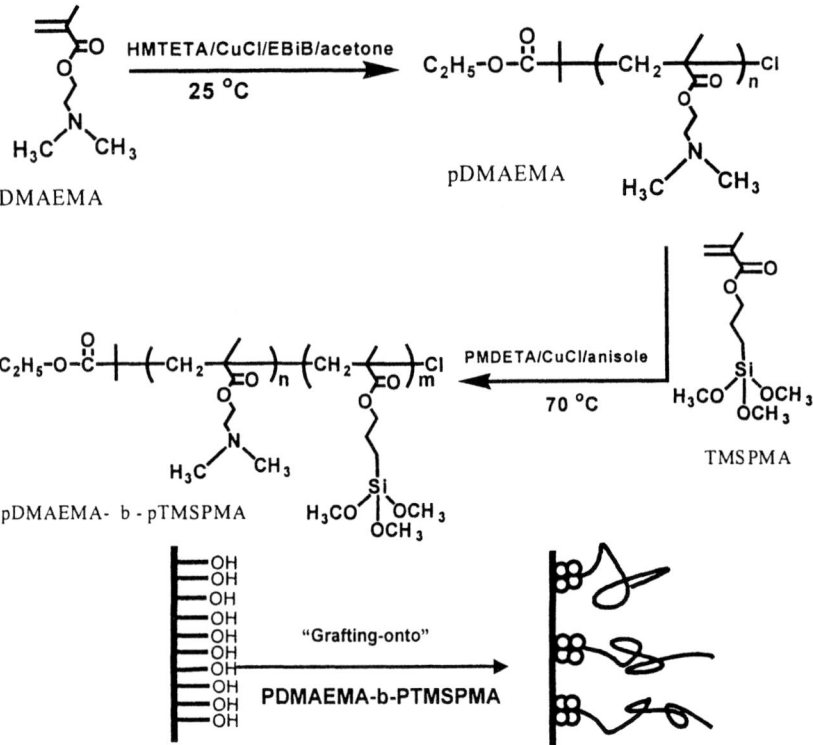

Scheme 1. Synthesis of pDMAEMA-b-pTMSPMA block copolymer in solution via ATRP and self-assembly to the surface of glass slide

pDMAEMA–b–pTMSPMA copolymer due to the temperature dependent solubility could change their conformation from extended to globule below and above LCST, respectively.

Block Copolymer Brushes on Flat Surfaces via "Grafting-from" Approach

The grafting-from approach, or grafts generated from a surface immobilized initiator, generally leads to the formation of a higher density of polymer brushes than the grafting-onto approach. By tuning parameters including grafting density, chemical composition or kind of applied surfaces, products with different properties could be achieved.

Brushes Generated from Oxidized Silicon Wafers Entirely Covered with Initiator.

Silicon wafers are among the most commonly used flat surfaces for these grafts. Initiators can be applied to both kinds of silicon substrates, namely oxidized (Si-OH) and hydrogen-terminated silicon (Si-H).[31,32] Grafting from the oxidized substrate via a chlorosilane (mono- or trichloro) functionalized initiator is the most documented route for grafting (Scheme 2).

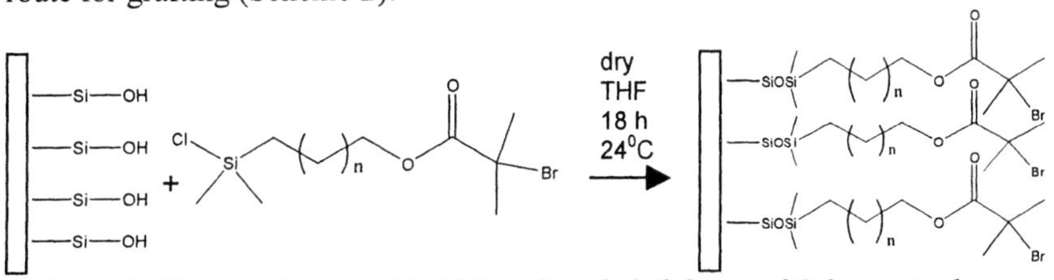

Scheme 2. The attachment of 1-(chlorodimethylsilyl)propyl 2-bromoisobutyrate as an ATRP initiator to silicon wafer.

The reaction occurs at room temperature over several hours. Formation of densely covered monolayers of initiator was confirmed by X-ray photon spectroscopy. In the absence of impurities deposited on the silicon wafers the only sources of carbon and bromine is an initiator layer.[31]

The preparation of block copolymer brushes using a "grafting-from" ATRP approach was first reported by tethering polystyrene-*block*-poly(*t*-butyl acrylate) (pSt-*b*-ptBA) to Si wafers.[31] Hydrolysis of the *t*-butyl groups yielded a polystyrene-*block*-poly(acrylic acid) brush demonstrating a versatile approach to tune film properties and wettability.

Water contact angles of a series of polymers prepared by ATRP from identically modified silicon wafers are presented in Table 1. In the first entry, a polystyrene layer 10 nm thick showed a contact angle of 90^0. When that surface was chain extended with additional 12 nm of poly(*t*-butyl acrylate), the surface became less hydrophobic (86°). Then, after hydrolysis of polyacrylate to poly(acrylic acid) thickness of the layer decreased to 16 nm due to relaxation of the chains upon removal of the bulky *t*-butyl groups. The presence of the acid was confirmed by the significant decrease of water contact angle from 86° to 18°. A very hydrophobic surface composed of poly(heptadecylfluorodecyl acrylate) was also constructed. The large contact angle of 119° is typical of surfaces containing high fluorine contents.

Table 1. Composition, thickness and water contact angles for films grown from silicon surfaces by ATRP.[31]

Polymer	thickness [nm]	contact angle [deg][a]
Polystyrene	12.4	90 ± 1
poly(styrene-b-*t*-butyl acrylate)	24.3	86 ± 4
poly(styrene-b-acrylic acid)	18.8	18 ± 2
poly(heptadecylfluorodecyl acrylate)	7.7	119 ± 2

[a] Determined by the horizontal plane method immediately after application of the water droplet.

Brushes Generated from Silicon Wafers with Varied Grafting Density.

Poly-(N,N-dimethylaminoethyl methacrylate) (pDMAEMA) was synthesized in the presence of a sacrificial initiator by surface-initiated ATRP from the silicon wafers with variable coverage of initiating groups. The concentration of initiating groups was varied by varying the molar ratio between the two chlorosilanes (active initiator vs. "dummy" initiator) used for the attachment through silanol groups on the surface of silicon wafers. For oxidized wafer surfaces, chlorotrimethylsilane was used as a "dummy" initiator and 1-(chlorodimethylsilyl)propyl 2-bromoisobutyrate as an active ATRP initiator, as shown in Scheme 3.

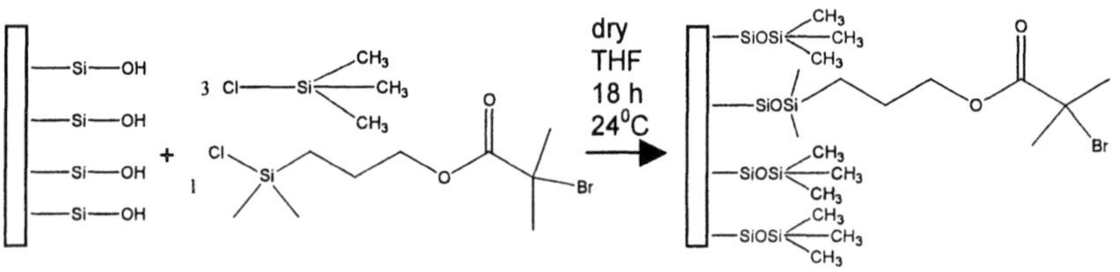

Scheme 3. The attachment of ATRP initiators groups and "dummy" initiator to the surface of silicon wafers.

Initiator grafting density was calculated under the assumption that the molar ratio of two chlorosilanes in solution corresponds to their molar ratio on the surface of silicon wafers.

Similar approach was used to attach the initiator to the Si-H substrates. Non-functionalized 1-alkene was used as a "dummy initiator" to vary the concentration of initiating groups on the hydrogen-passivated s ilicon s ubstrate (Scheme 4). To attached initiating groups on silicon substrates freshly hydrogen-passivated silicon wafers were exposed to UV light in the presence of 1- alkene and a llyloxytrimethylsilane (TMS) for a hydrosilylation reaction. In a second step, the TMS group was converted to 2-bromoisobutyrate ester (an ATRP initiator) using the corresponding acid bromide in the presence of tetrabutylammonium fluoride in dry tetrahydrofuran (Scheme 4).

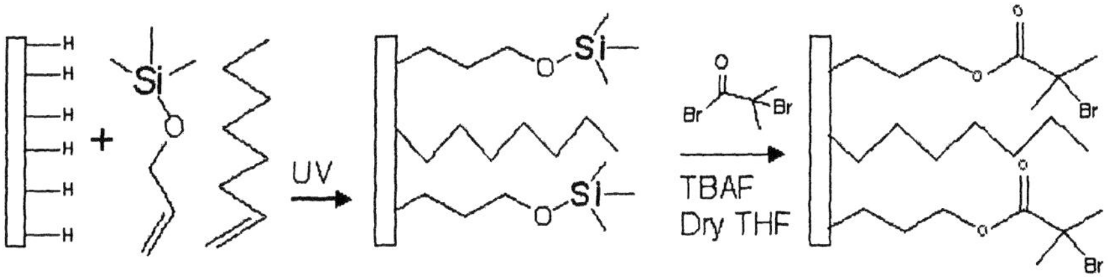

Scheme 4. The attachment of ATRP initiators groups and "dummy" initiator to the surface of hydrogen-terminated silicon wafers.

Due to extremely low amount of initiator attached to the silicon wafers, quantitative c haracterization o f t he d egree o f s urface f unctionalization poses a particular challenge. Neither ellipsometry nor XPS can be used to provide the reliable measure of homogeneity of surface coverage. However it is possible to determine the surface coverage when initiating sites are "amplified" by the polymer chains built from the initiator. Since polymer brushes are inherently larger than the initiator molecules, increase in surface roughness is usually observed.[33] Sacrificial initiator added at the beginning of the reaction allows controlled chains growth. It is needed because the low concentrations of initiating groups cannot generate a sufficient concentration of persistent radical (deactivator).[34-36] Alternatively, excess of persistent radical (Cu^{II} species) can be initially added.[2]

Tapping mode AFM images of pDMAEMA grafts prepared using modified Si-OH surface and imaged in air under ambient conditions are shown in Figure 2 a-c. Images of grafts obtained at higher then 20% initiator coverage (Figure 2 a) revealed the presence of dense, relatively smooth films, with occasional patchy defects.

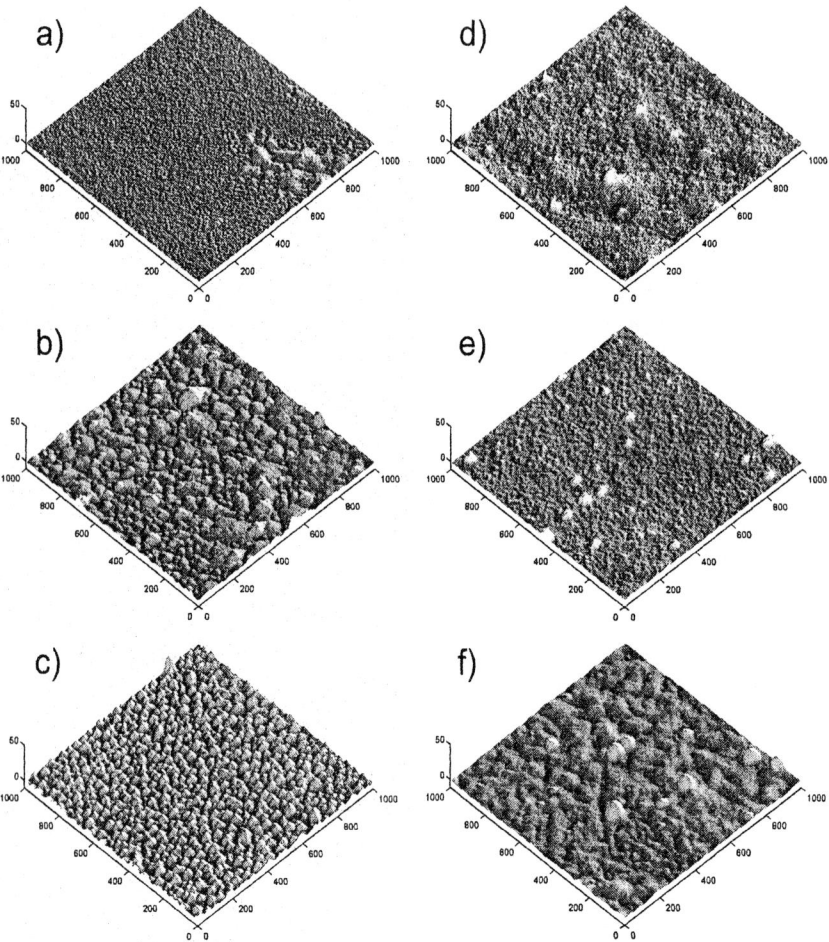

Figure 2. AFM images of Si-OH (left images) vs. Si-H (right images) substrates covered with pDMAEMA layer synthesized by ATRP: a) 20%, $M_n = 95\ 000$; $M_w/M_n = 1.24$, b) 13 %, $M_n = 78\ 000$, $M_w/M_n = 1.26$, c) 1%, $M_n = 78\ 000$; $M_w/M_n = 1.26$ and for Si-H: d) 20%, $M_n = 78\ 000$; $M_w/M_n = 1.4$, e) 13%, $M_n = 78\ 000$; $M_wM_n = 1.4$, f) 1%, $M_n = 78\ 000$; $M_w/M_n = 1.26$. (All scales in nm).

Upon lowering the initiator coverage, the grafts were increasingly patchy, and at 1% coverage and below, they had a form of highly uniform, isolated patches. The average apparent volumes of these patches (not corrected for the finite size of the AFM tip) was on the order of 2000 nm^3, whereas their average height measured with respect to the exposed areas of the wafer did not exceed 10 nm. Such dimensions of patches indicate that they consisted of just a few individual polymer chains, and that the chains were collapsed onto the surrounding surface.

AFM images of grafts grown from the Si-H substrates are shown in Figure 2 (d-f). It appears that, in comparison with oxidized silicon substrates with the same assumed initiator coverage, films grown from Si-H substrates were denser, showing fragmentation into patches only at the lowest initiator coverage (Fig. 2 f). The use of both substrates (Si-OH vs. Si-H) seems to yield equally uniform graft layers when the concentration of initiating group on the surface is ≥20%, and does not appear to offer any clear advantages for ATRP synthesis.

Temperature responsiveness of similar pDMAEMA grafts on gold surfaces was studied in water by surface plasmon resonance. Gold surfaces were prepared using a disulfide initiator (bis(2-ethyl-2-bromoisobutyrate) disulfide) dissolved in acetone with dibutyl disulfide as the non-initiating group. Polymerization of pDMAEMA occurred in the glove box with the use of sacrificial initiator. Upon heating in water above the LCST, thickness of pDMAEMA films synthesized by ATRP from gold surfaces with 20% initiator coverage decreased from 41 nm to 33 nm (pDMAEMA synthesized in solution M_n= 93 000, M_n/M_w= 1.4).

Molecular Bottle Brushes Generated from Silicon Wafers with Varied Grafting Density.

The synthesis of stimulus-responsive molecular bottle brushes using grafting-from approach started by generating polymer backbone attached to the surface. Poly(2-hydroxyethyl methacrylate) (pHEMA) was first grown from the oxidized silicon wafers. The pendent hydroxyl groups in pHEMA brushes were converted to 2-bromoisobutyrate esters (pBIEM) using standard organic methods.[37] In a method analogous to preparation of molecular bottle brushes in solution, the pBIEM chains attached at one chain end to the surface were used to grow poly(N,N-dimethylaminoethyl methacrylate) side chains (Scheme 5).

Tapping mode AFM images of pHEMA grafts are shown in Figure 3. As in the previously discussed case of linear chains of pDMAEMA attached to the silicon wafer, with the decrease of initiator coverage, the grafts exhibited the increase of "patchiness". It was accompanied by the decrease of individual patch size, down to the volume comparable with the total volume of just a few polymer chains.

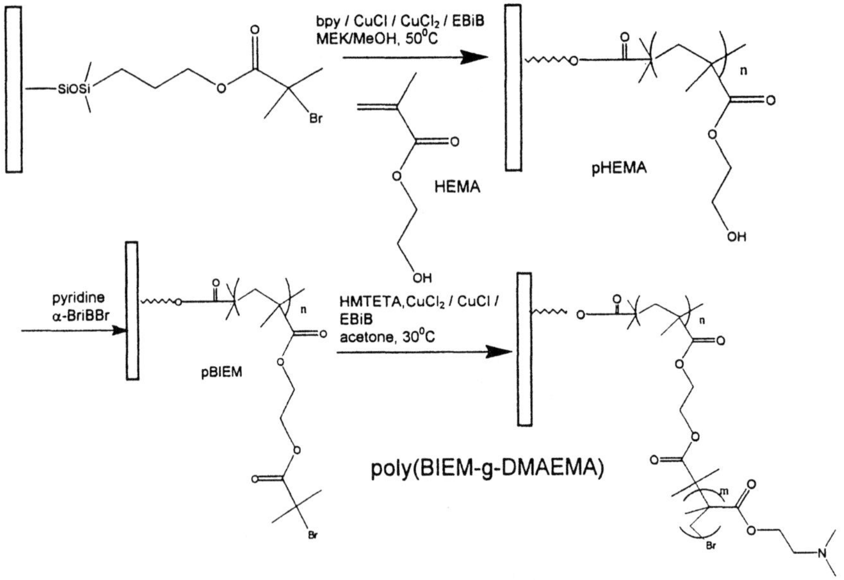

Scheme 5. Synthesis of poly[(2-(isobutyryloxy)ethyl methacrylate)-g- (N,N-dimethylaminoethyl methacrylate)] brush attached to silicon wafers.

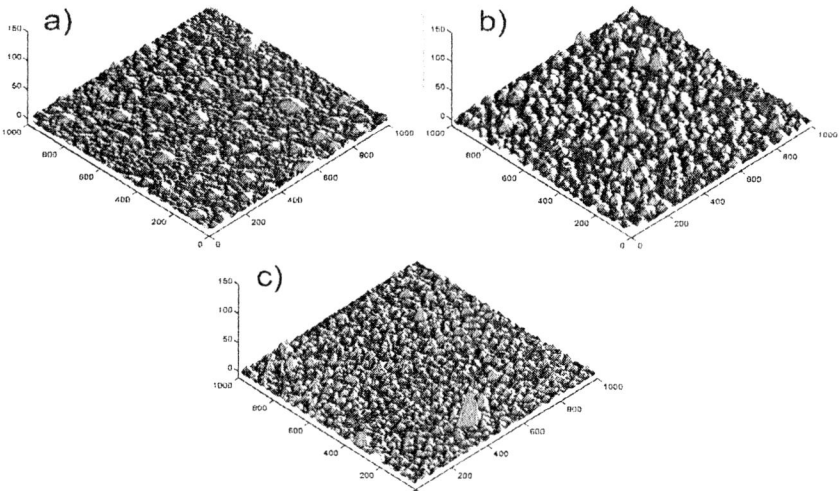

Figure 3. AFM images of PHEMA synthesized from silicon wafers with variable density of initiator (f) a) f=9%, M_n = 64 000; M_w/M_n = 1.1, b) f=1%, M_n = 64 000; M_w/M_n = 1.2, c) f=0.1%, M_n = 68 000, M_w/M_n = 1.2. (All scales in nm).

Table 3. Molecular weights and polydispersities of poly(N,N-dimethylaminoethyl methacrylate) synthesized in the presence of sacrificial initiator and calculated molecular weight of poly[(2-(isobutyryloxy)ethyl methacrylate)-g- (N,N-dimethylaminoethyl methacrylate)] brushes attached to silicon wafers.

Initiator coverage f	DP pHEMA	M_n pDMAEMA (side chains)	M_w/M_n pDMAEMA (side chains)	$M_{n, theor}$ poly(BIEM-g-DMAEMA (brush)
50 %	490	5 300	1.06	2 600 000
9%	490	5 300	1.07	2 600 000
1%	490	5 800	1.11	2 800 000
0.1%	520	6 000	1.09	3 100 000

Molecular weights and polydispersities of pDMAEMA synthesized in the presence of sacrificial initiator, shown in Table 2, demonstrate that synthesis of side chains preceded with good control ($M_w/M_n \leq 1.1$). Molecular weight of poly(BIEM-g-DMAEMA) brushes synthesized from pBIEM attached to silicon wafers was calculated from equation $M_{n,\,theor} = DP_{HEMA} \times M_{n\,pDMAEMA}$.

AFM images of short pHEMA grafts (5% initiator coverage, $M_n = 14\,000$, $M_w/M_n = 1.1$) acquired before and after attachment of pDMAEMA side chains ($M_{n\,pDMAEMA} = 5\,700$, $M_w/M_n = 1.07$, $M_{n\,theor\,poly(BEMA\text{-}gDMAEMA)} = 627\,000$) are shown in Figure 4 a and b, respectively. The image of short pHEMA "backone" graft resembles the images of longer pHEMA grafts (Figure 3). Successful attachment of pDMAEMA side chains was manifested by increase of "bulkiness" of individual patches. (Figure 4)

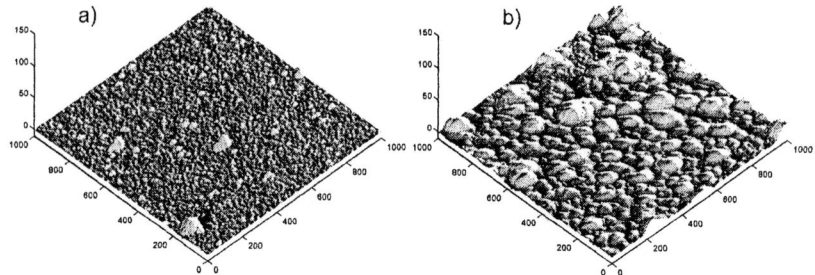

Figure 4. AFM images of a) pHEMA b) poly(BIEM-g-DMAEMA) brush synthesized from silicon wafers with 5% of initiator coverage, (a) $M_n = 14\,000$; $M_w/M_n = 1.1$, b) $M_n = 627\,000$). (All scales in nm).

Summary

Attachment of thin polymer films was successfully used for surface property modification. The grafting onto/from approaches were used to immobilize polymer layers on silicon and gold substrates depending on the desired polymer density. ATRP was demonstrated to be a versatile technique to synthesize well-defined polymer brushes. The grafting density of the attached brushes was varied by changing the molar ratio of ATRP active initiator and "dummy" initiator attached to surfaces. In the case of very low initiator coverage, isolated polymer domains consisting of a few individual chains were obtained. When higher initiator coverage was used, uniform polymer layer was observed ($\geq 20\%$). Molecular bottle brushes were successfully synthesized from flat silicon wafers. Surface morphology of polymer films was varied by treatment with different solvents and temperatures.

Acknowledgements

The authors would like to thank the National Science Foundation (grant 03-04568), CRP Consortium at Carnegie Mellon University and the Kosciuszko Foundation for financial support.

References

(1) Advincula, R.; Brittain, W. J.; Caster, K. C.; Ruhe, J. *Polymer Brushes*; 2004 WILEY-VCH Verlag GmbH & Co. KGaA: Weinheim, 2004.
(2) Pyun, J.; Kowalewski, T.; Matyjaszewski, K. *Macromol. Rapid Comm.* **2003**, *24*, 1043-1059.
(3) Halperin, A.; Tirrell, M.; Lodge, T. P. *Adv. Polym. Sci.* **1992**, *100*, 31-71.
(4) Lodge, T. P.; Muthukumar, M. *J. Phys. Chem.* **1996**, *100*, 13275-13292.
(5) Edmondson, S.; Huck, W. T. S. *J. Mater.Chem.y* **2004**, *14*, 730-734.
(6) Wang, J.-S.; Matyjaszewski, K. *J. Am. Chem. Soc.* **1995**, *117*, 5614-5615.
(7) Patten, T. E.; Xia, J.; Abernathy, T.; Matyjaszewski, K. *Science* **1996**, *272*, 866-868.
(8) Patten, T. E.; Matyjaszewski, K. *Acc. Chem. Res.* **1999**, *32*, 895-903.
(9) Matyjaszewski, K. *Chem.--Eur. J.* **1999**, *5*, 3095-3102.
(10) Coessens, V.; Pintauer, T.; Matyjaszewski, K. *Prog. Polym. Sci.* **2001**, *26*, 337-377.
(11) Matyjaszewski, K.; Xia, J. *Chem. Rev.* **2001**, *101*, 2921-2990.
(12) Kamigaito, M.; Ando, T.; Sawamoto, M. *Chem. Rev.* **2001**, *101*, 3689-3745.
(13) Husseman, M.; Malmstroem, E. E.; McNamara, M.; Mate, M.; Mecerreyes, D.; Benoit, D. G.; Hedrick, J. L.; Mansky, P.; Huang, E.; Russell, T. P.; Hawker, C. J. *Macromolecules* **1999**, *32*, 1424-1431.
(14) Matyjaszewski, K.; Miller, P. J.; Pyun, J.; Kickelbick, G.; Diamanti, S. *Macromolecules* **1999**, *32*, 6526-6535.
(15) Zhao, B.; Brittain, W. J.; Zhou, W.; Cheng, S. Z. D. *J. Am. Chem. Soc.* **2000**, *122*, 2407-2408.
(16) von Werne, T. A.; Germack, D. S.; Hagberg, E. C.; Sheares, V. V.; Hawker, C. J.; Carter, K. R. *J. Am. Chem. Soc.* **2003**, *125*, 3831-3838.
(17) Boyes, S. G.; Brittain, W. J.; Weng, X.; Cheng, S. Z. D. *Polym. Preprints* **2002**, *43*, 549-550.
(18) Kong, X.; Kawai, T.; Abe, J.; Iyoda, T. *Macromolecules* **2001**, *34*, 1837-1844.
(19) Zhao, B. *Polymer* **2003**, *44*, 4079-4083.
(20) Ionov, L., Sidorenko, A., Stamm, M., Minko, S., Zdyrko, B., Klep, V., Luzinov, I. *Macromolecules* **2004**, *37*, 7421 - 7423.

(21) Edmondson, S.; Osborne, V. L.; Huck, W. T. S. *Chem. Soc. Rev.* **2004**, *33*, 14-22.
(22) Matyjaszewski, K.; Qin, S.; Boyce, J. R.; Shirvanyants, D.; Sheiko, S. S. *Macromolecules* **2003**, *36*, 1843-1849.
(23) Kowalewski, T.; McCullough, R. D.; Matyjaszewski, K. *Eur. Phys. J. E: Soft Matter* **2003**, *10*, 5-16.
(24) Tsubokawa, N.; Yoshikawa, S. *Recent Res. Dev. Polym. Sci.* **1998**, *2*, 211-228.
(25) von Werne, T.; Patten, T. E. *J. Am. Chem. Soc.* **1999**, *121*, 7409-7410.
(26) von Werne, T.; Patten, T. E. *J. Am. Chem. Soc.* **2001**, *123*, 7497-7505.
(27) Huang, W.; Baker, G. L.; Bruening, M. L. *Angew. Chem., Int. Ed.* **2001**, *40*, 1510-1512.
(28) Mori, H.; Seng, D. C.; Zhang, M.; Mueller, A. H. E. *Langmuir* **2002**, *18*, 3682-3693.
(29) Zhao, B.; Brittain, W. J. *Prog. Polym. Sci.* **2000**, *25*, 677-710.
(30) Pyun, J.; Jia, S.; Kowalewski, T.; Matyjaszewski, K. *Macromol. Chem. Phys.* **2004**, *205*, 411-417.
(31) Matyjaszewski, K.; Miller, P. J.; Shukla, N.; Immaraporn, B.; Gelman, A.; Luokala, B. B.; Siclovan, T. M.; Kickelbick, G.; Vallant, T.; Hoffmann, H.; Pakula, T. *Macromolecules* **1999**, *32*, 8716-8724.
(32) Yu, W. H.; Kang, E. T.; Neoh, K. G.; Zhu, S. *J. Phys. Chem. B* **2003**, *107*, 10198-10205.
(33) Zhao, B.; Brittain, W. J.; Zhou, W.; Cheng, S. Z. D. *Macromolecules* **2000**, *33*, 8821-8827.
(34) Ejaz, M.; Yamamoto, S.; Ohno, K.; Tsujii, Y.; Fukuda, T. *Macromolecules* **1998**, *31*, 5934-5936.
(35) Goto, A.; Fukuda, T. *Prog. Polym. Sci.* **2004**, *29*, 329-385.
(36) Fischer, H. *Chem. Rev.* **2001**, *101*, 3581-3610.
(37) Cheng, G.; Boeker, A.; Zhang, M.; Krausch, G.; Mueller, A. H. E. *Macromolecules* **2001**, *34*, 6883-6888.

Chapter 3

Synthesis of Terminally Functionalized (Co)Polymers via Reversible Addition Fragmentation Chain Transfer Polymerization and Subsequent Immobilization to Solid Surfaces with Potential Biosensor Applications

Charles W. Scales[1], Anthony J. Convertine[1], Brent S. Sumerlin[1], Andrew B. Lowe[2,*], and Charles L. McCormick[1,2,*]

Departments of [1]Polymer Science and [2]Chemistry and Biochemistry, The University of Southern Mississippi, Hattiesburg, MS 39406

Herein we report our recent progress in the preparation of stimuli-responsive, terminally functionalized (co)polymers by a controlled radical polymerization (CRP) technique known as reversible addition fragmentation chain transfer (RAFT) polymerization. We also investigate the potential application of RAFT-generated polymers for sensor and microfluidic applications and the preparation of polymeric bio-conjugates. Three classes of polymers—salt-repsonsive, pH-responsive, and temperature-responsive—have been synthesized in our laboratories and our attempts to immobilize these on to surfaces are reviewed.

© 2005 American Chemical Society

Introduction

Our group has a long-standing interest in the synthesis and characterization of water-soluble (co)polymers. In more recent times, our work has focused on the tailored synthesis of pH, temperature, and salt-responsive homo- and block-copolymers in both organic and aqueous media *via* RAFT polymerization (*1-6*). The RAFT methodology is of great synthetic utility in that it allows for the polymerization of several monomer classes including, but not limited to, anionic, cationic, zwitterionic, and neutral systems. This allows for the direct preparation of stimuli-responsive (co)polymers of various polymer architectures, including diblocks, stars, and combs. Additionally, the inherent nature of RAFT allows for the synthesis of polymers having reactive chain-end functionality.

Synthesis of Electrolyte-Responsive Polymers *via* RAFT

Prior to our work, only a few examples of the direct synthesis of controlled-structure, near-monodisperse poly(betaines) had been reported. Lowe, Billingham, and Armes (*7-9*) utilized group transfer polymerization to prepare near-monodisperse poly(2-dimethylamino)-ethyl methacrylate homopolymers which were then reacted with 1,3 propanesultone. In related work, Jaeger *et al.* (*10, 11*) synthesized poly(4-vinylpyridine) by nitroxide-mediated polymerization and subsequently derivatized the precursor (co)polymer to yield the corresponding carboxy- or sulfobetaine. More recently, Lobb *et al.* (*12*) reported the direct polymerization of 2-methacryloyloxyethyl phosphorylcholine in aqueous media *via* atom transfer radical polymerization (ATRP) (*13*). The first examples of controlled molecular weight polybetaines synthesized by RAFT were those reported by Donovan, Lowe, and McCormick, (14) in which poly(*N*-(2-*N,N*-dimethylaminoethyl)-*N*-acrylamide) (M_n=9300 and M_w/M_n=1.23) was homo- and copolymerized using benzyldithiobenzoate as the RAFT agent. The homopolymer was then derivatized with 1,3-propanesultone to yield the corresponding poly(sulfopropylbetaine). The first direct, controlled synthesis of poly(sulfobetaines) *via* RAFT was reported by Donovan *et al.* (*2*). and involved the RAFT polymerization of styrenic, acrylamido and methacrylate-based sulfobetaine-containing monomers, (Figure 1) including 3-[2-(*N*-methylacrylamido)-ethyldimethylammonio]propanesulfonate (MAEDAPS), 3-[*N*-(2-methacroyloyethyl)-*N,N*-dimethylammonio]propanesulfonate (DMAPS), and 3-(*N,N*-dimethylvinylbenzylammonio)propanesulfonate (DMVBAPS).

All three monomers were polymerized in *aqueous* salt solution (0.5 M NaBr) at 70 °C using sodium 4-cyanopentanoic acid dithiobenzoate (CTPNa) as the RAFT agent and 4,4′-azobis(4-cyanopentanoic acid) (V-501) as the initiating species. Molecular weight and PDI vs. conversion and the kinetic profile for MAEDAPS are shown in Figure 2.

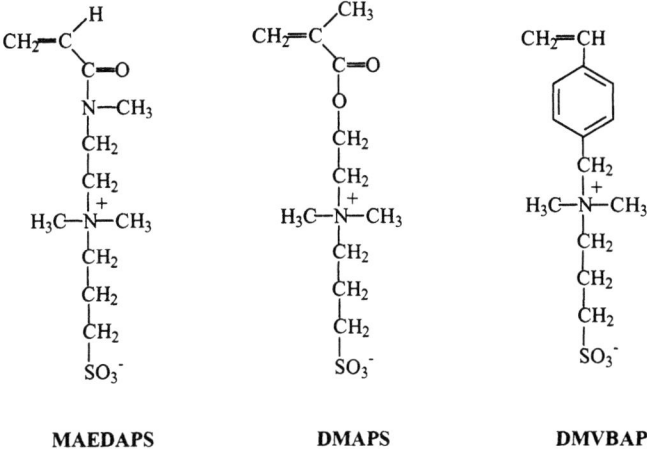

Figure 1. Chemical structures of MAEDAPS, DMAPS, and DMVBAPS

Synthesis of pH-Responsive Polymers *via* RAFT

The synthesis of water-soluble, pH responsive systems *via* RAFT has also been of great interest to our group. The first examples of this by Mitsukami *et al.* (5) were the separate CTPNa-mediated RAFT block copolymerizations of sodium 4-styrenesulfonate (4-SS) and (*ar*-vinylbenzyl)trimethylammonium chloride (VBTAC) with the pH-responsive sodium 4-vinyl benzoate (4VB) and *N,N*-dimethylvinylbenzylamine (DMVBA), respectively. These block copolymers contain a permanently hydrophilic block and a potentially hydrophobic, pH-responsive block. The ability to reversibly form polymeric micelles as a function of solution pH was demonstrated utilizing dynamic light scattering. Sumerlin and coworkers (*1*) further demonstrated the ability of RAFT to polymerize the industrially significant, pH responsive, sodium 3-acrylamido-3-methylbutanoate (AMBA) monomer. This macro-CTA was reacted with sodium 2-acrylamido-2-methylpropanesulfonate (AMPS), forming a pH-responsive block-copolymer capable of undergoing reversible micelle formation. More recently, Convertine *et al.* (*15*) employed cumyl dithiobenzoate (CDB) as the RAFT agent in the separate homopolymerizations of 2 and 4-vinylpyridine (2VP and 4VP). Block copolymers composed of both monomers were also easily prepared (Figures 3 and 4).

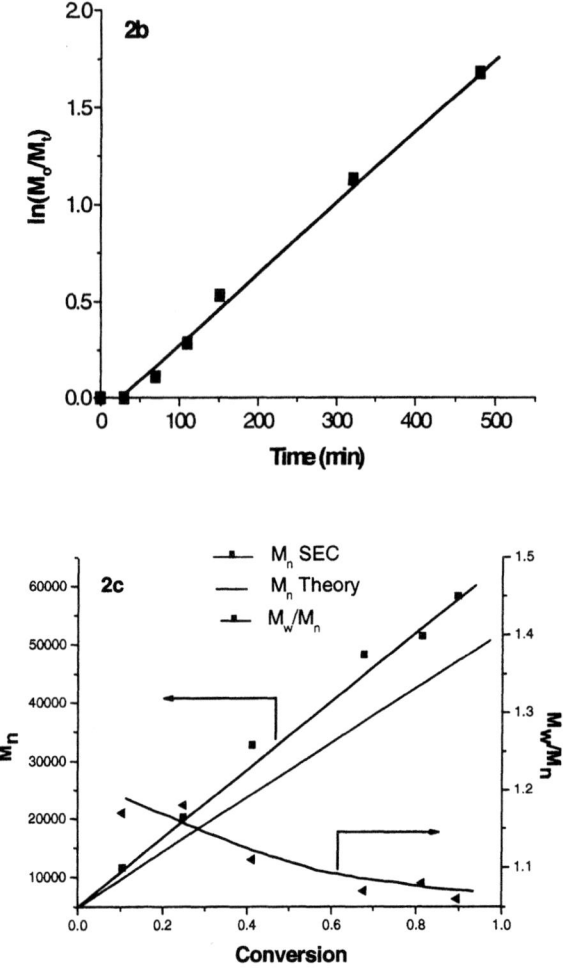

Figure 2. Controlled polymerization of MAEDAPS showing the pseudo-first-order kinetic plot and the corresponding M_n vs conversion data.

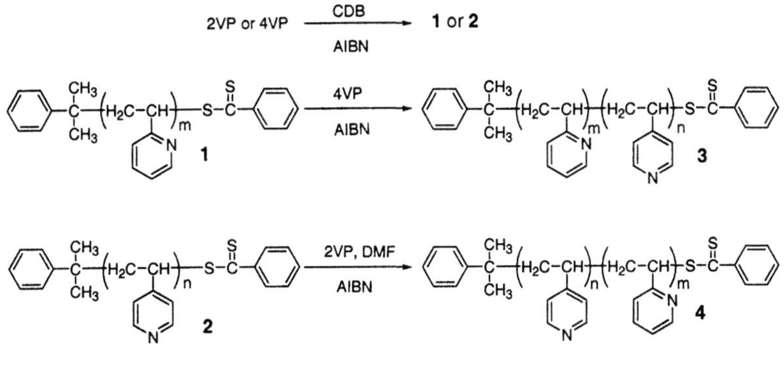

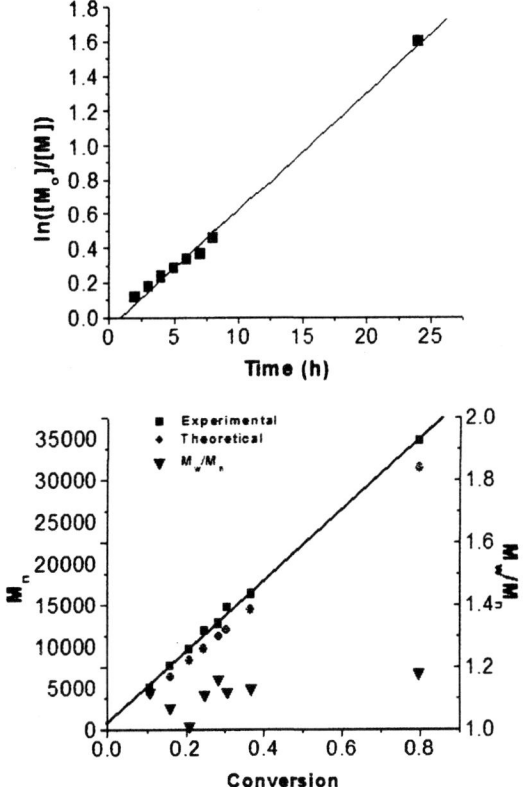

Figure 3. Synthetic pathways for the polymerization of poly(2VP) and poly(4VP) and block copolymers of 2VP and 4VP at 60 °C via RAFT.

Figure 4. Controlled polymerization of 2VP showing the pseudo-first-order kinetic plot and the corresponding M_n vs conversion data.

Vasilieva et al. (16) also synthesized in aqueous media, a pH responsive system based on the cationic, methacrylamido monomer N-[3-(dimethylamino) propyl]methacrylamide (DMAPMA) utilizing 4-cyanopentanoic acid dithiobenzoate (CTP) (Figure 5). Polymerization in water at neutral pH allowed a moderate level of control over the polymerization up to 50% conversion. On the other hand, polymerization in an aqueous buffer (pH = 5.0) afforded excellent control up to 98% conversion M_n = 38,000 g/mole, M_w/M_n = 1.12 (Figure 6).

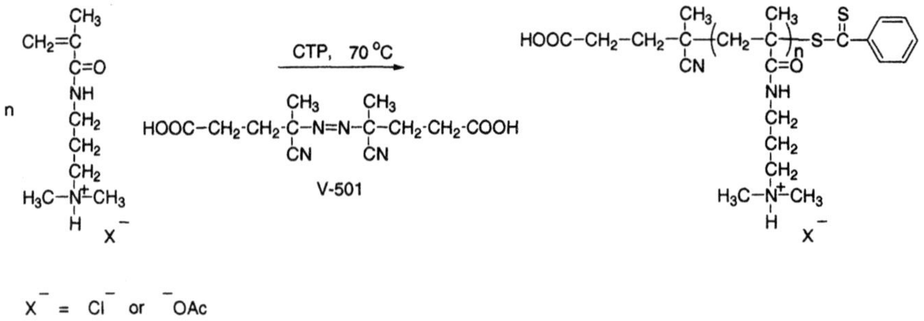

Figure 5. Overall scheme for the RAFT polymerization of DMAPMA

Synthesis of Temperature-Responsive Polymers *via* RAFT

Most recently, Convertine et al. (17) reported the facile, controlled, room-temperature RAFT polymerization of N-isopropylacrylamide (NIPAAM) in N,N-dimethylformamide (DMF) using the conventional azo-initiator, 2,2'-azobis(4-methoxy-2,4-dimethylvaleronitrile) (V-70) as a primary radical source and the commercially available, *trithiocarbonate*-based CTA, 2-dodecylsulfanyl-thiocarbonylsulfanyl-2-methyl propionic acid (DMP) (Figures 7 and 8).

NIPAAM is an important nonionic acrylamido monomer that has been the subject of intense research over the years due to the readily accessible lower critical solution temperature (LCST) of its homopolymer in water of ~32 °C. The proximity to human body temperature (37 °C) offers opportunities for drug delivery applications (18-21). At present, there are several reports detailing the RAFT polymerization of NIPAAM. For example, Ganachaud et al. (22) reported the AIBN initiated solution polymerization of NIPAAM employing both benzyl dithiobenzoate (in benzene) and cumyl dithiobenzoate (in 1,4-dioxane) at 60 °C.

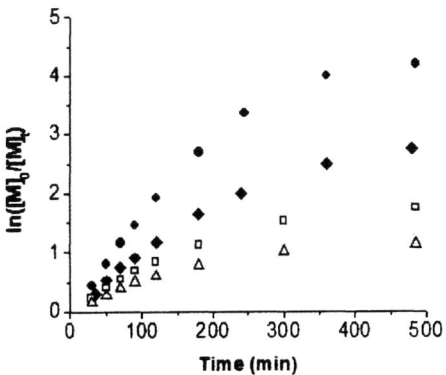

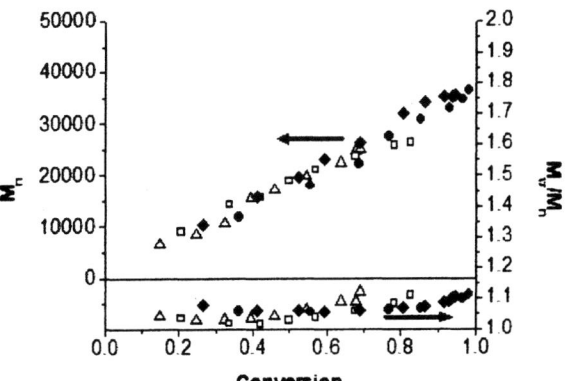

Figure 6. Controlled polymerization of DMAPMA showing the pseudo-first-order kinetic plots: [CTA]/[I]) 1.5/1 (circles), [CTA]/[I]) 3/1 (diamonds), [CTA]/[I]) 5/1 (squares), and [CTA]/[I]) 8/1 (triangles) and the dependence of M_n and M_w/M_n on the monomer conversion.

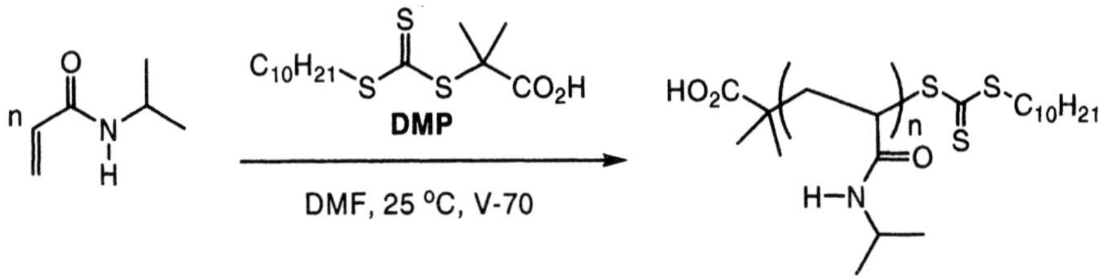

Figure 7. Synthetic pathway for the room temperature RAFT polymerization of N-isopropylacrylamide.

These authors also conducted a thorough size exclusion chromatographic (SEC) investigation of the resulting homopolymers and highlighted some of the problems associated with the analysis of PNIPAAM samples using this technique. Subsequently, Schilli *et al.* disclosed the benzyl and cumyl dithiocarbamate-mediated polymerization of NIPAAM, also in 1,4-dioxane at 60 °C (*23*). These experimental conditions led to polymers with polydispersity indices (PDIs) around 1.37. Polymer molecular weights were determined by a combination of MALDI-TOF MS and SEC. The problems associated with the SEC analysis of PNIPAAM were noted with experimentally determined molecular weights being considerably higher than those theoretically predicted. More recently, Ray and co-workers demonstrated the ability to control the tacticity in RAFT polymerizations of NIPAAM *via* the addition of a suitable Lewis acid such as $Sc(OTf)_3$ or $Y(OTf)_3$ (*24, 25*). Using an eluent comprised of DMF containing 0.1 M LiCl, the authors did not report any of the difficulties in GPC analysis of the polymers previously reported. RAFT-prepared PNIPAAM has also been employed as a thermoresponsive stabilizing layer for gold nanoparticles/clusters (*26, 27*).

Application of RAFT-Generated Polymers

The versatility of polymers synthesized *via* RAFT allows many possible applications in areas such as pharmaceutics, bio-sensing, and biotechnology. Our laboratory recently reported the facile immobilization of such polymers onto gold colloids and films. This method employs an *in situ* process in which the dithioester chain-end is reduced to the corresponding thiol in the presence of an unconjugated metal such as gold. Chemisorption of the polymeric thiols allows for formation of polymerically-stabilized nanoparticles in solution (*28*) or direct immobilization onto gold films (*29*) (Figure 9). Reaction conditions were optimized for specific polymer systems, including those with anionic, cationic and neutral pendant functionality. Currently we are assessing the ability of these polymers to extend their functional chain-ends into solution allowing for further

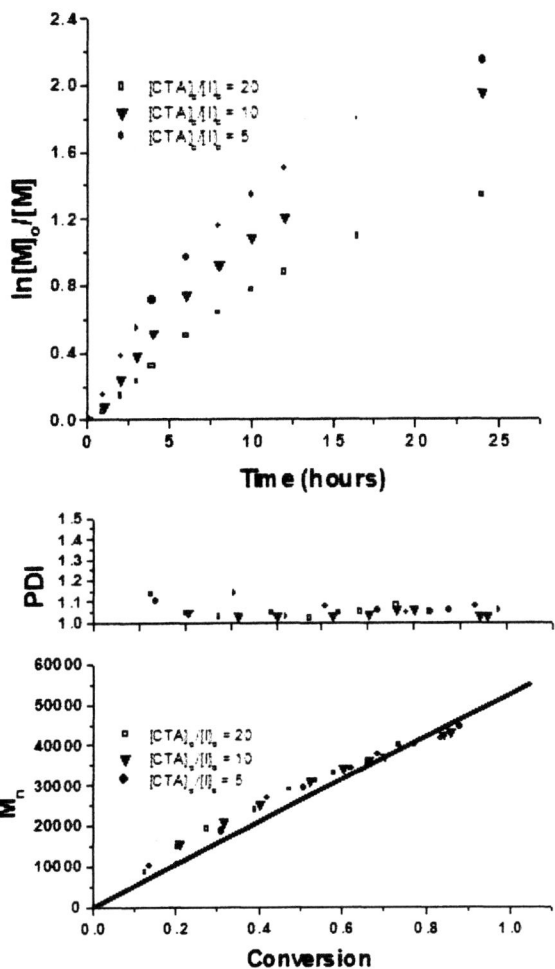

Figure 8. Controlled polymerization of NIPAAM showing the pseudo first-order rate plots for the NIPAAM homopolymerizations as a function of variation in $[CTA]_0/[I]_0$ and plots of PDI and M_n versus conversion for a PNIPAAM homopolymerization at multiple $[CTA]_0/[I]_0$ ratios.

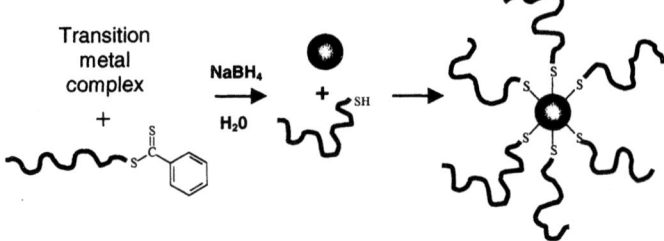

Figure 9. Preparation of (co)polymer-stabilized transition metal nanoparticles

chemical modification or attachment of bio-ligands including peptides or oligonucleotides.

Experimental

Monomer and Polymer Syntheses

The syntheses of MAEDAPS, DMAPS, DMVBAPS (*2*) and sodium 4-vinyl benzoate (*5*) as well as all RAFT agents employed, have been described in detail in earlier publications (*6*). Other commercially available monomers such as, AMPS, AMBA, 4-SS, DMVBA, VBTAC, 2VP, 4VP and NIPAAM were purchased from Sigma-Aldrich and used as described in their respective publications (*1-3, 5, 14-17*).

Characterization Techniques and Instrumentation

All molecular weight data was obtained using Size Exclusion Chromatography (SEC) coupled with multi-angle laser light scattering (MALLS). Two SEC-MALLS systems were employed, including an aqueous-based Wyatt 18-angle DAWN-EOS (with an OptilabDSP Interferometric Refractometer ($\lambda = 690$ nm)) and an organic-based Viscotek TDA 302 (with RI, viscosity, 7 mW 90° and 7° true low angle light scattering detectors, ($\lambda = 670$ nm)). The specific SEC conditions for each polymer system have been described in earlier publications (*1-3, 5, 14-17*).

Conclusions

In this chapter, we have highlighted some recently reported examples of the synthesis of water-soluble, stimuli-responsive polymers *via* RAFT. Additionally, we discussed possible applications for these multifunctional, telechelic polymer systems, mentioning specifically their use in the stabilization of transition-metal nanoparticles and immobilization onto metal films *via* their thiol terminus. The telechelic nature of these systems, along with their response to specific changes in solution properties (*i.e.* salt concentration, pH, and temperature) make them attractive candidates for use in microfluidics, biosensing, and pharmaceutics.

Acknowledgements

Major support for these studies from the National Science Foundation Materials Research Science Engineering Center (DMR 0213883), the Department of Energy, and Genzyme Pharmaceuticals, is gratefully acknowledged.

References

(1) Sumerlin, B. S.; Donovan, M. S.; Mitsukami, Y.; Lowe, A. B.; McCormick, C. L. *Macromolecules* **2001**, *34*, 6561.
(2) Donovan, M. S.; Sumerlin, B. S.; Lowe, A. B.; McCormick, C. L *Macromolecules* **2002**, *35*, 8663.
(3) Donovan, M. S.; Sanford, T.; Lowe, A. B.; Sumerlin, B. S.; Mitsukami, Y.; McCormick, C. L *Macromolecules* **2002**, *35*, 4570.
(4) Lowe, A. B.; McCormick, C. L. *Aus. J. Chem.* **2002**, *55*, 367.
(5) Mitsukami, Y.; Donovan, M. S.; Lowe, A. B.; McCormick, C. L. *Macromolecules* **2001**, *34*, 2248.
(6) McCormick, C. L.; Lowe, A. B. *Acc. Chem. Res.* **2004**, *37*, 312.
(7) Lowe, A. B.; Billingham, N. C.; Armes, S. P. *Chem. Commum.* **1996**, 1555.
(8) Butun, V.; Bennett, C. E.; Vamvakaki, M.; Lowe, A. B.; Billingham, N. C.; Armes, S. P. *J. Mat. Chem.* **1997**, *7*, 1693.
(9) Lowe, A. B.; Billingham, N. C.; Armes, S. P. *Macromolecules* **1999**, *32*, 2141.
(10) Bohrisch, J.; Schimmel, T.; Englehardt, H.; Jaeger, W. *Macromolecules* **2002**, *35*, 4143.
(11) Jaeger, W.; Wendler, U.; Lieske, A.; Bohrisch, J.; Wandrey, C. *Macromol. Symp.* **2000**, *161*.
(12) Lobb, E. J.; Ma, I.; Billingham, N. C.; Armes, S. P.; Lewis, A. L. *J. Am. Chem. Soc.* **2001**, *123*, 7913.
(13) Li, Y.; Armes, S. P.; Jin, X.; Zhu, S. *Macromolecules* **2003**, *36*, 8268.
(14) Donovan, M. S.; Lowe, A. B.; McCormick, C. L *Poly. Prepr.* **1999**, *40*, 281.
(15) Convertine, A. J.; Sumerlin, B. S.; Thomas, D. B.; Lowe, A. B.; McCormick, C. L. *Macromolecules* **2003**, *36*, 4679.

(16) Vasilieva, Y. A.; Thomas, D. B.; Scales, C. W.; McCormick, C. L. *Macromolecules* **2004**, *37*, 2728.
(17) Convertine, A. J.; Ayres, N.; Scales, C. W.; Lowe, A. B.; McCormick, C. L. *Biomacromolecules* **2004**, *5*, 1177.
(18) Dube, D.; Francis, M.; Lerous, J. C.; Winnik, F. M. *Bioconjugate Chemistry* **2002**, *13*, 685.
(19) Eeckman, F.; MoNs, A. J.; Amighi, K.; *Int. J. Pharm.* **2004**, *273*, 101.
(20) Eeckman, F.; MoNs, A. J.; Amighi, K.; *Eur. Polym. J.* **2004**, *40*, 873.
(21) Yamazaki, A.; Winnik, F. M.; Cornelius, R. M.; Brash, J. L. *Biophys. Biochim. Acta* **1999**, *1421*, 103.
(22) Ganachaud, F.; Montiero, M. J.; Gilbert, R. G.; Dourges, M. A.; Thang, S. H.; Rizzardo, E. *Macromolecules* **2000**, *33*, 6738.
(23) Schilli, C.; Lanzendoerfer, M. G.; Mueller, A. H. E. *Macromolecules* **2002**, *35*, 6819.
(24) Ray, B.; Isobe, Y.; Matsumoto, K.; Habaue, S.; Okamoto, Y.; Kamigaito, M.; Sawamoto, M. *Macromolecules* **2004**, *37*, 1702.
(25) Ray, B.; Isobe, Y.; Morioka, K.; Habaue, S.; Okamoto, Y.; Kamigaito, M.; Sawamoto, M. *Macromolecules* **2003**, *36*, 543.
(26) Raula, J.; Shan, J.; Nuopponen, M.; Niskanen, A.; Jiang, H.; Kauppinen, E. I.; Tenhu, H. *Langmuir* **2003**, *19*, 3499.
(27) Shan, J.; Nuopponen, M.; Jiang, H.; Kauppinen, E.; Tenhu, H. *Macromolecules* **2003**, *36*, 4526.
(28) Lowe, A. B.; Sumerlin, B. S.; Thomas, D. B.; McCormick, C. L. *J. Am. Chem. Soc.* **2002**, *124*, 11562.
(29) Sumerlin, B. S.; Lowe, A. B.; Stroud, P. A.; Zhang, P.; Urban, M. W.; McCormick, C. L. *Langmuir* **2002**, *19*, 5559.

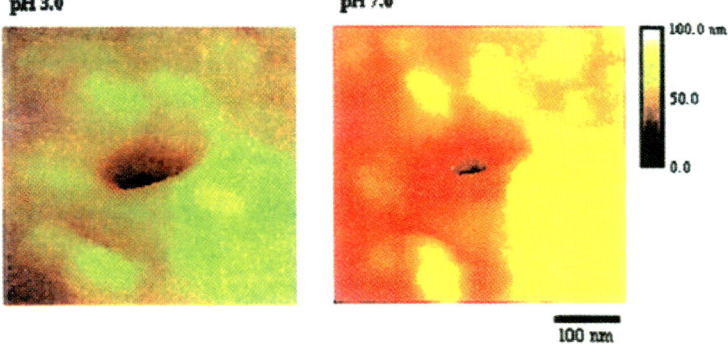

Plate 1.7. Self-assembled porous membranes at pH 3.0 and pH 7.0 (Reproduced from reference 99. Copyright 2000 American Chemical Society.)

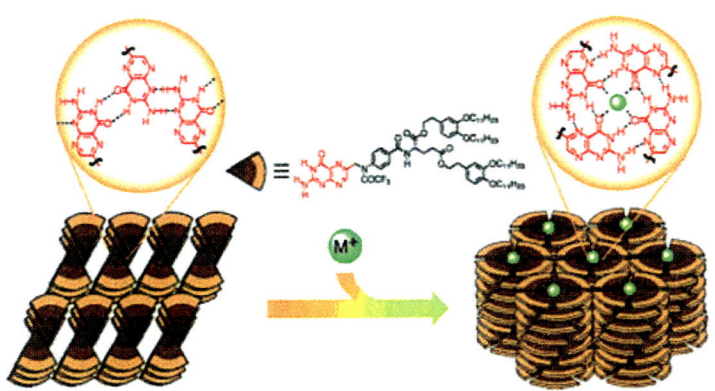

Plate 1.9. Ion-responsive self-assembled LC crystals with ribbon-like structures of a folic acid derivative and their structural change to the disc-like tetramers are obtained by the addition of metal ions (M^+) (Reproduced with permission from reference 167. Copyright 2002 American Association for the Advancement of Science.)

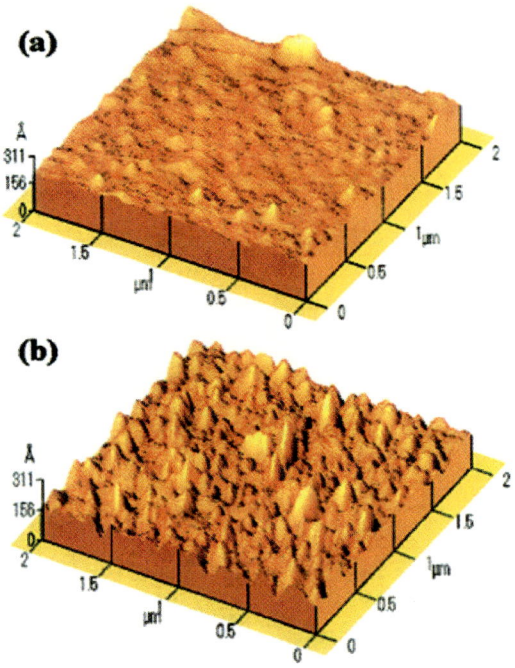

Plate 4.6. AFM images of the Si/SiO$_2$//PS-b-PAA(Ag$^+$) polymer brush (a) before and (b) after reduction

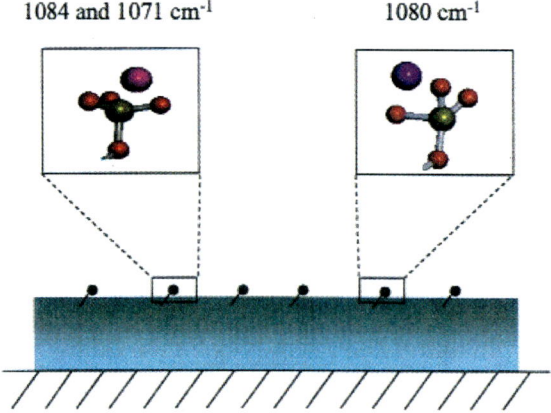

Plate 8.2. Preferential orientation of –SO$_3^-$Na$^+$ groups near the surface of Sty/EHA/MAA (Adapted from reference 15. Copyright 2003 American Chemical Society.)

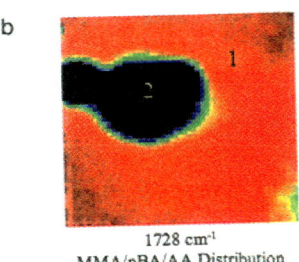

Plate 8.5. b – IRIRI image obtained by tuning into the 1728 cm^{-1} IR band after 24 hours of coalescence. (Reproduced from reference 65. Copyright 2003 American Chemical Society.)

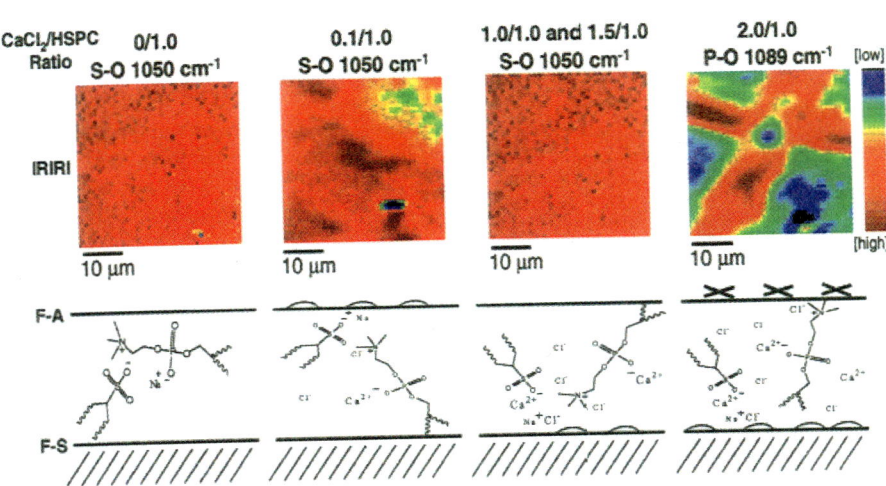

Plate 8.6. Schematic representation of particle interactions: A – without Ca^{2+}; B and C – in the presence of Ca^{2+} ions and the effect of $CaCl_2$/HSPC ratio on mobility and crystallization at the F-A interface (Reproduced from reference 34. Copyright 2004 American Chemical Society.)

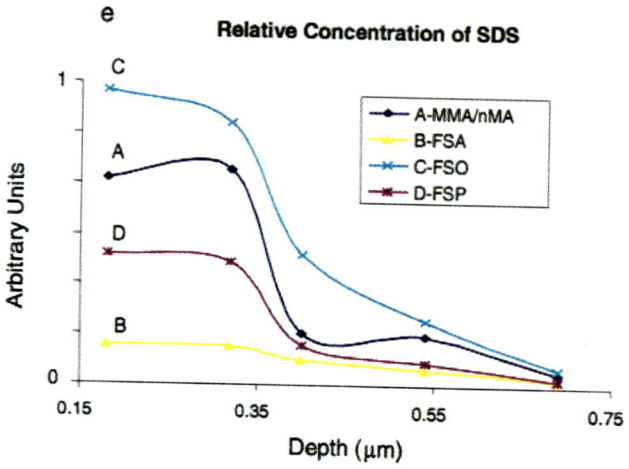

Plate 8.7.e – Relative concentration levels of SDS plotted as a function of depth from the F-A interface for: A – MMA/nBA containing FSA; B – MMA/nBA containing FSO; and C – MMA/nBA containing FSP (Reproduced from reference 66. Copyright 2004 American Chemical Society.)

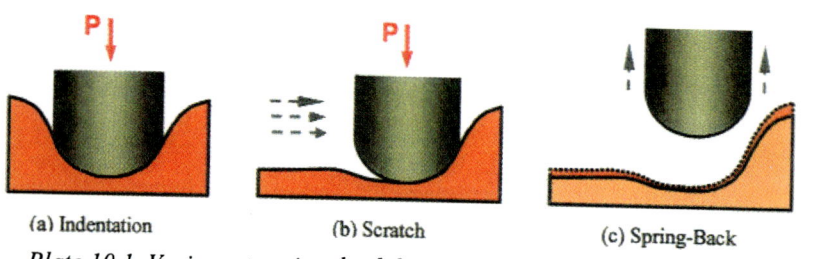

Plate 10.1. Various steps involved during a scratch process (load-controlled).

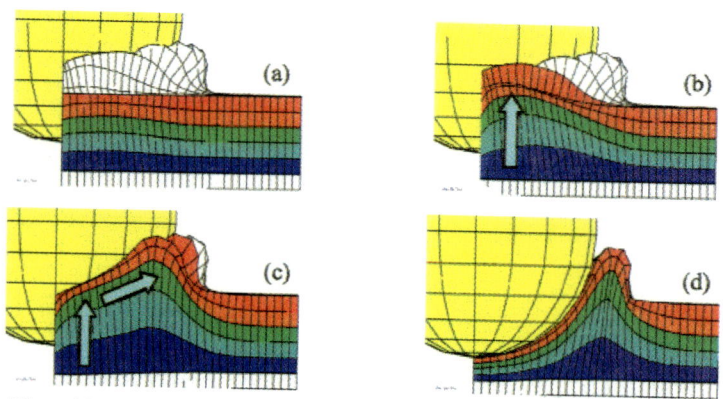

Plate 10.4: (a) Undeformed section; (b) section is compressed and squeezed upwards; (c) section is pushed to the side; (d) a scratch groove is formed.

Chapter 4

Synthesis and Application of Polyelectrolyte Brushes

Stephen G. Boyes[1], Crystal Cyrus[2], Bulent Akgun[2], Adam Caplan[2], Brian Mirous[2], and William J. Brittain[2,*]

[1]School of Polymers and High Performance Materials, The University of Southern Mississippi, Hattiesburg, MS 39406–0076
[2]Department of Polymer Science, The University of Akron, Akron, OH 44325–3909

The synthesis of tethered polyelectrolyte brushes by the use of atom transfer radical polymerization (ATRP) offers a versatile technique with many advantages over traditional free radical polymerization methods. This group has synthesized both homopolymer and diblock copolymer polyelectrolyte brushes using ATRP. The response of these systems to changes in solvent and salt concentration has been investigated, with the degree of response being dependent upon the brush composition. The use of diblock copolymer polyelectrolyte brushes for synthesis of metallic nanoparticles has also been demonstrated.

Surface modification with thin polymer films has been widely used to modify surface properties such as biocompatibility, corrosion resistance, and wetability. One such surface modification method that has recently attracted considerable attention is polymer brushes *(1-6)*. Polymer brushes refer to an assembly of polymer chains that are tethered by one end to a surface or interface *(7,8)*. Tethering of the chains in close proximity to each other forces the chains to stretch away from the surface to avoid overlapping. Thin polymer films are typically synthesized by either physisorption or covalent attachment. Of these methods, covalent attachment is preferred due to enhanced thermal and solvolytic stability when compared to physisorption techniques *(2)*. Covalent attachment of polymer chains can be achieved using either the "grafting to" or "grafting from" techniques. The "grafting to" technique involves tethering preformed end-functionalized polymer chains to substrates via reaction between the chain end and the surface *(9)*. The "grafting from" technique involves the immobilizing of initiators onto the substrate, followed by in situ surface-initiated polymerization to generate the tethered polymer brush. The "grafting from" approach has generally become the most attractive method to prepare thick, covalently tethered polymer brushes with a high grafting density *(2)*. It is our belief that only the "grafting from" technique is capable of producing a sufficiently high grafting density so that the chains can be considered polymer brushes.

Recent advances in polymer synthesis techniques have given rise to the importance of controlled/"living" free radical polymerization, as it provides a number of advantages over traditional free radical techniques *(10)*. The main advantages provided by a controlled/"living" free radical system for polymer brush synthesis are control over brush thickness and uniformity, via control of molecular weight and narrow polydispersities, and the ability to prepare block copolymers by sequential activation of the dormant chain end in the presence of different monomers *(5,11)*. Probably the most common controlled/"living" free radical polymerization technique used to produce polymer brushes is atom transfer radical polymerization (ATRP) *(5,11-13)*.

Recently there has been increasing interest in the preparation, characterization, and application of polyelectrolyte brushes *(14-16)*. Surfaces functionalized with polyelectrolyte brushes may have important applications in fields such as "smart" coatings, biosensors, and colloid stabilization. Particular interest is in systems where the swelling of the tethered polyelectrolyte brush can be prompted by an external stimulus such as the pH, temperature, or ionic strength of the surrounding environment *(16)*.

This chapter presents a brief overview of the authors' findings in the synthesis of polyelectrolyte brushes, tethered to flat surfaces, using ATRP. The stimuli responsive behavior of these systems and an application in the synthesis of metallic nanoparticle hybrid systems is also discussed.

Synthesis of Homopolymer Polyelectrolyte Brushes

The synthesis of a tethered homopolymer brush of quaternized poly(*N,N*-dimethylaminoethyl methacrylate) (PDMAEMA) on a flat silicon substrate via an immobilized ATRP initiator is outlined in Figure 1. Preparation of the homopolymer polyelectrolyte brush involved use of the "grafting from" approach to produce a homopolymer brush of PDMAEMA followed by treatment with iodomethane to produce the quaternized salt of PDMAEMA. Treatment of the homopolymer PDMAEMA brush with iodomethane resulted in an increase in the brush thickness from 15 nm to 26 nm and an increase in the advancing water contact angle from 60° to 71°. The attenuated total reflectance Fourier transform infrared spectroscopy (ATR-FTIR) spectra for the PDMAEMA brush contained characteristic peaks for the symmetric stretching vibration of the methyl group in the $-N(CH_3)_2$ moiety (2820 and 2770 cm^{-1}) and carbonyl group (1735 cm^{-1}). Treatment of the PDMAEMA brush with iodomethane resulted in the loss of the peaks at 2820 and 2770 cm^{-1} due to the formation of the quaternized salt.

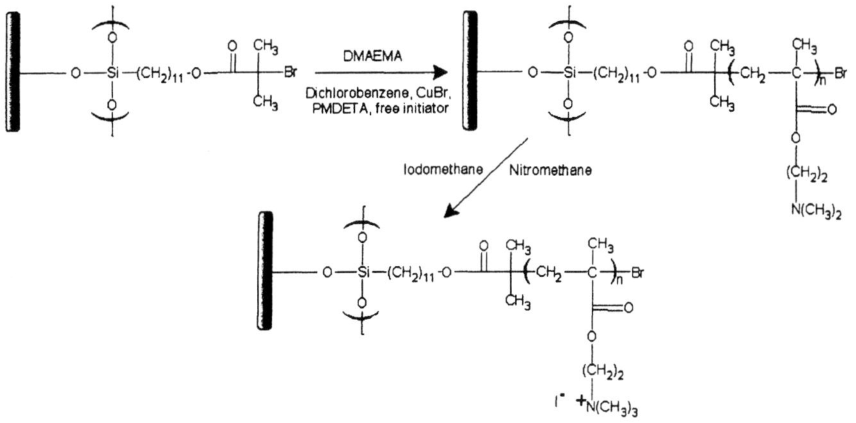

Figure 1. Synthesis of surface-immobilized polyelectrolyte homopolymer brush (Si/SiO$_2$//PDMAEMA) using atom transfer radical polymerization.

In order to examine the stimuli-responsive behavior of the homopolymer brushes, the quaternized PDMAEMA polymer brush was treated with a 1 M solution of potassium iodide. The height of the polymer brush should be influenced by the salt concentration of the surrounding environment due to electrostatic interactions within the polymer brush *(14)*. As such, the thickness of the quaternized PDMAEMA brush was measured before and after treatment with the potassium iodide solution for 1 h. It should be noted that the thickness measurements were conducted on dried samples, which may not accurately reflect the equilibrium state, but general trends should still be observed *(14)*. Ellipsometric measurements indicated that the thickness of the quaternized PDMAEMA brush decreased by 4 nm after treatment with the salt solution. This decrease is not as large as that reported by Osborne *et al.*, who observed decreases up to 17 nm when homopolymer brushes of poly(2-(methacryloyloxy)ethyl)trimethylammonium chloride) (PMETAC) were treated with 1 M potassium iodide solutions *(14)*. Further work is currently being conducted to examine the stimuli responsive nature of the quaternized PDMAEMA polymer brush systems.

Synthesis of Polyelectrolyte Diblock Copolymer Brushes

The synthesis of tethered polyelectrolyte diblock copolymer brushes of either polystyrene (PS) or poly(methyl acrylate) (PMA) and poly(acrylic acid) (PAA) on flat silicate substrates via the "grafting from" approach is outlined in Figure 2 *(17)*. Preparation of the polyelectrolyte brushes involves immobilization of the ATRP initiator, (11-(2-bromo-2-methyl)propionyloxy)undecyl trichlorosilane, onto a silicon substrate by self-assembly techniques followed by sequential ATRP of either styrene or methyl acrylate and *tert*-butyl acrylate to produce the desired diblock copolymer brush. The diblock copolymer brush, containing the *tert*-butyl acrylate block, was then hydrolyzed to the acrylic acid by refluxing the brush in a solution of 10% aqueous HCl for 1 – 12 h. Shorter refluxing times were used for samples when PMA was the tethered block, as longer refluxing times resulted in degrafting of the polymer chains from the surface. After hydrolysis, the diblock copolymer brushes containing the PAA blocks were treated with 10 mM of either a silver or palladium salt to form the polyelectrolyte diblock copolymer brush.

Table I outlines the properties for each of the polyelectrolyte diblock copolymer brushes synthesized. Each of the values shown is representative of the outer layer only. The results in Table 1 indicate that after hydrolysis of the *tert*-butyl acrylate, regardless of the tethering block, the thickness decreases by approximately 50% and there is a dramatic decrease in the water contact angle of the surface. The decrease in thickness after hydrolysis is believed to be as a result of chain relaxation caused by removal of the bulky *tert*-butyl groups and has been reported by other researchers *(5,18)*. The dramatic decrease in advancing contact angle is believed to be due to the formation of the acrylic acid, as PAA should produce a surface that is completely wettable. Treatment of the polyelectrolyte diblock copolymer brushes containing a PAA layer with metal salts, either silver acetate or sodium tetrachloropalladate, resulted in an increase in the thickness of the outer block and in the advancing water contact angle. These increases were attributed to attachment of metal cations to the PA block.

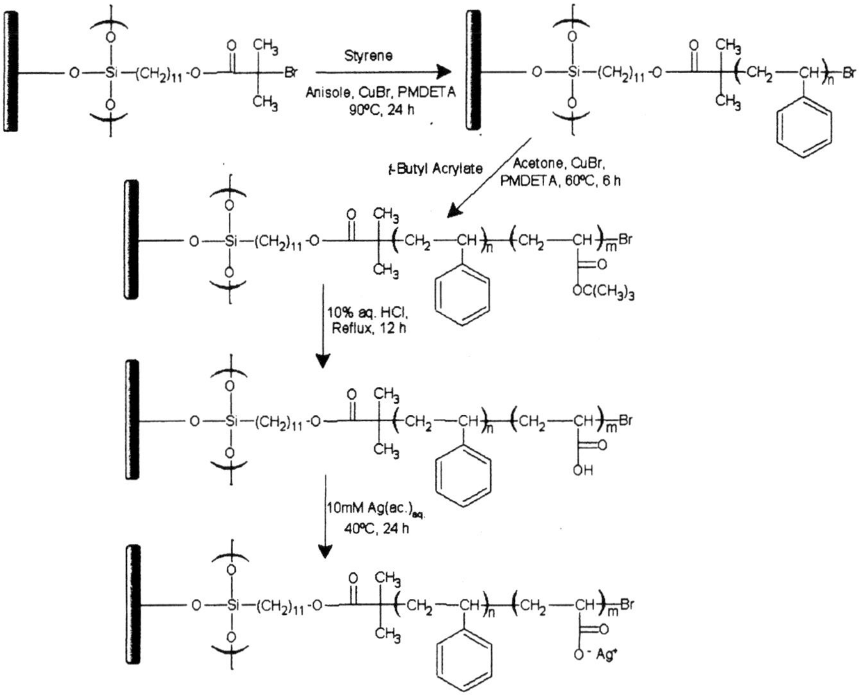

Figure 2. Synthesis of surface-immobilized polyelectrolyte diblock copolymer brush (Si/SiO$_2$//PS-b-PAA(Ag$^+$)) using atom transfer radical polymerization (17).

The ATR-FTIR spectra for the synthesis of the tethered Si/SiO$_2$//PS-*b*-PAA(Ag$^+$) polymer brush are shown in Figure 3. The initial PS layer (Figure 3a) showed characteristic peaks at 3026, 3060, and 3084 cm^{-1}, due to aromatic C-H stretching vibrations, and at 1453 and 1493 cm^{-1}, due to the aromatic C-C stretching vibrations. There are also both symmetric and asymmetric stretching vibrations due to the backbone CH$_2$ groups at 2850 and 2925 cm^{-1}, respectively. Addition of the P(*t*-BA) layer (Figure 3b) resulted in the presence of a peak at 1733 cm^{-1} due to the C=O stretch, a peak at 2976 cm^{-1} due to the asymmetric CH$_3$ stretching vibration, and a doublet at 1367 and 1392 cm^{-1} from the symmetric methyl deformation of the *tert*-butyl group. Hydrolysis of the P(*t*-BA) layer to form PAA (Figure 3c) was confirmed by the presence of a broad OH stretch at 2900 – 3400 cm^{-1}, a broadening and slight downward shift of the C=O peak to 1710 cm^{-1}, and loss of both the asymmetric CH$_3$ stretch and symmetric methyl deformation doublet due to removal of the *tert*-butyl groups. Treatment of the PAA layer with an aqueous metal salt (Figure 3d depicts the reaction product of silver acetate) resulted in loss of the broad OH peak and a shift in the C=O peak at 1710 cm^{-1} to 1548 cm^{-1} which was attributed to the formation of the carboxylate anion. Similar results were seen in the ATR-FTIR spectra for the synthesis of the Si/SiO$_2$//PS-*b*-PAA(Pd^{2+}) polymer brush. Similar results were also seen for the Si/SiO$_2$//PMA-*b*-PAA(Ag$^+$) polymer brush except characteristic peaks for PMA instead of PS were present.

Table I. Physical Properties of Polyelectrolyte Diblock Copolymer Brushes

Brush Structure	Water Contact Angle^a (deg)		Thickness^b (nm)
	Θ_a	Θ_r	
Si/SiO$_2$//PS	99	84	21
Si/SiO$_2$//PS-*b*-P(*t*-BA)	95	84	17
Si/SiO$_2$//PS-*b*-PAA	Complete wetting		8
Si/SiO$_2$//PS-*b*-PAA(Ag$^+$)	35	18	11
Si/SiO$_2$//PS	100	86	21
Si/SiO$_2$//PS-*b*-P(*t*-BA)	94	76	12
Si/SiO$_2$//PS-*b*-PAA	Complete wetting		5
Si/SiO$_2$//PS-*b*-PAA(Pd^{2+})	26	9	11
Si/SiO$_2$//PMA	81	70	14
Si/SiO$_2$//PMA-*b*-P(*t*-BA)	95	82	16
Si/SiO$_2$//PMA-*b*-PAA	30	14	9
Si/SiO$_2$//PMA-*b*-PAA(Ag$^+$)	40	25	13

^{a} The standard deviation of contact angles was < 2°. ^{b} Thickness determined by ellipsometry and is representative of the outer block only. Typical error on thickness measurement is ± 1 nm.

X-ray photoelectron spectroscopy (XPS) was used to characterize the samples both before and after treatment with the aqueous metal salts, and in each case demonstrated the successful attachment of metal cations to the diblock copolymer brush. In the case of the Si/SiO$_2$//PS-*b*-PAA polymer brush, treatment with silver acetate resulted in the presence of 9.6 atomic% of Ag$^+$ in the XPS spectra and treatment with sodium tetrachloropalladate resulted in 6.6 atomic% of Pd^{2+}. When the Si/SiO$_2$//PMA-*b*-PAA polymer brush was treated with silver acetate, the XPS spectra showed the presence of 10.2 atomic% of Ag$^+$.

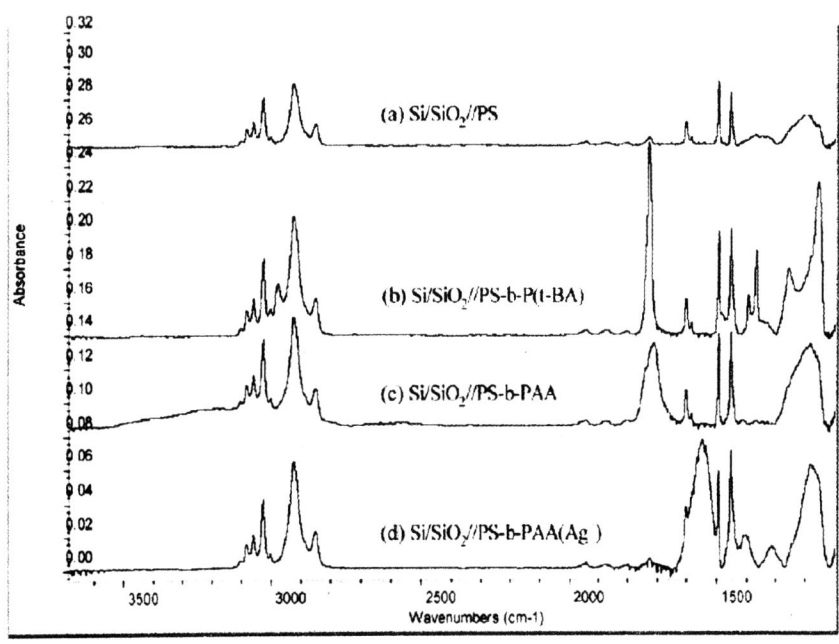

Figure 3. ATR-FTIR Spectra for the preparation of the Si/SiO$_2$//PS-b-PAA(Ag$^+$) polymer brush.

Solvent Treatment of Polyelectrolyte Diblock Copolymer Brushes

Block copolymer brushes are currently one of the most interesting architectures produced due to the vertical phase separation that occurs when the block copolymer chains are tethered by one end to a surface or substrate. It has been predicted theoretically *(1)* and, in some cases, demonstrated experimentally *(11,19,20)*, that block copolymer brushes can form a variety of novel well-ordered structures by changing factors such as grafting density, chain length, block composition, or the interaction energy between the blocks and the surrounding environment. To investigate the ability of polyelectrolyte diblock copolymer brushes to undergo surface rearrangement, selective solvent treatment of the samples was conducted.

The Si/SiO$_2$//PS-*b*-PAA and Si/SiO$_2$//PMA-*b*-PAA polymer brushes, both before and after treatment with silver acetate, were treated with a solvent that was a non-solvent for the outer block but a good solvent for the tethered block. For each of the diblock copolymer brushes, N,N-dimethylformamide (DMF) was used to extend the brush because DMF is a good solvent for PS, PMA, and PAA. After treatment with DMF, the diblock copolymer brushes were treated with anisole, which is a good solvent for both PS and PMA but a non-solvent for PAA in both the non-salt and salt states. Table II contains the advancing water contact angle results for the solvent treatment of the Si/SiO$_2$//PS-b-PAA polymer brush. Similar results were seen for the Si/SiO$_2$//PMA-b-PAA polymer brush. After treatment with DMF, the advancing water contact angle was low (< 40°) due to the presence of either PAA or PAA(Ag$^+$) at the interface of the brush. Subsequent treatment with anisole resulted in an increase of the advancing water contact angle for each system. However, the values obtained after treatment with anisole were not characteristic of the advancing water contact angles of PS (≈ 100°) or PMA (≈ 76°). The maximum advancing water contact angle obtained after treatment with anisole was 48°, indicating that surface rearrangement of the diblock copolymer brush was very limited. Previous reports have indicated that surface rearrangement is strongly dependent upon the Flory-Huggins interaction parameter between the two blocks (20). In this case, it appears as though the interaction parameter between either PS or PMA and the PAA, in either the salt or non-salt form, is high enough to inhibit complete surface rearrangement of the diblock copolymer brushes.

Table II. Selective Solvent Treatment of Si/SiO$_2$//PS-b-PAA Polymer Brush Both Before and After Treatment with Silver Acetate

Solvent Treatment[a]	Si/SiO$_2$//PS-b-PAA		Si/SiO$_2$//PS-b-PAA(Ag$^+$)	
	Θ_a	Θ_r	Θ_a	Θ_r
1st DMF	24	9	44	35
1st Anisole	48	28	48	41
2nd DMF	26	12	44	39
2nd Anisole	47	30	48	40

[a] Solvent treatment was performed at 40 °C for 30 mins.

Formation of Metal Nanoparticles from Polyelectrolyte Polymer Brushes

Nanocomposites of inorganic nanoparticles embedded within a polymer matrix have recently attracted a great deal of interest due to their potential application in areas such as catalysis, magnetic information storage, photonics, and biosensors (21-23). Much of this interest stems from the fact that nanometer-sized inorganic nanoparticles demonstrate unique properties, primarily due to quantum confinement effects and their large surface area relative to volume (24). As an example, polyelectrolyte multilayers have been previously reported as suitable polymeric template for the synthesis of metallic nanoparticles (23,25). This technique relies on the synthesis of polyelectrolyte multilayer films with a controlled content of free carboxylic acid binding groups. These binding groups can then be used to attach various metal cations, which can subsequently be reduced to form metallic nanoparticles. To examine the suitability of polyelectrolyte diblock copolymer brushes to produce metallic nanoparticles, the $Si/SiO_2//PS$-b-$PAA(Ag^+)$ and $Si/SiO_2//PS$-b-$PAA(Pd^{2+})$ polymer brushes were both reduced using H_2 (2 atm) (Figure 4).

The ATR-FTIR spectra for the preparation of silver nanoparticles from the $Si/SiO_2//PS$-b-$PAA(Ag^+)$ polymer brush are shown in Figure 5. The $Si/SiO_2//PS$-b-$PAA(Ag^+)$ brush was reduced in the presence of H_2 (2atm) at 120 °C for 2 days. Comparison of the ATR-FTIR spectra before and after reduction (Figure 5b and c) show the re-formation of the broad OH peak at 2900-3400 cm^{-1} and a shift from of the carboxylate anion peak at 1548 cm^{-1} to a carbonyl peak at 1710 cm^{-1}. These results indicate that after the reduction reaction the PAA block shifts from the ionized form to the protonated form, which suggests that the attached silver ions are being reduced to zerovalent silver nanoparticles. Similar results were seen in the ATR-FTIR spectra for reduction of the polyelectrolyte diblock copolymer brushes that contained palladium cations.

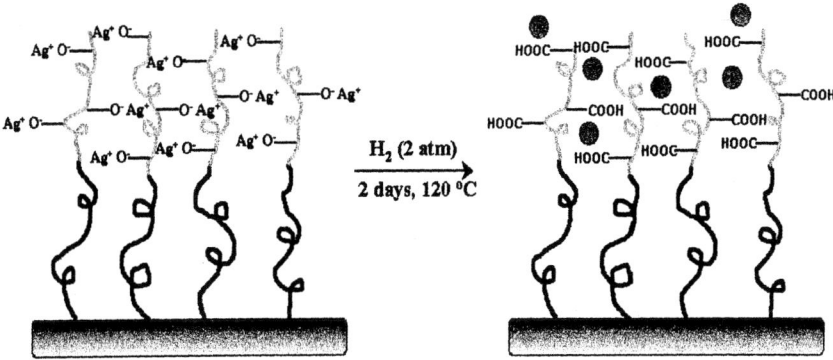

Figure 4. Reduction of polyelectrolyte diblock copolymer brush containing silver cations to produce silver nanoparticles

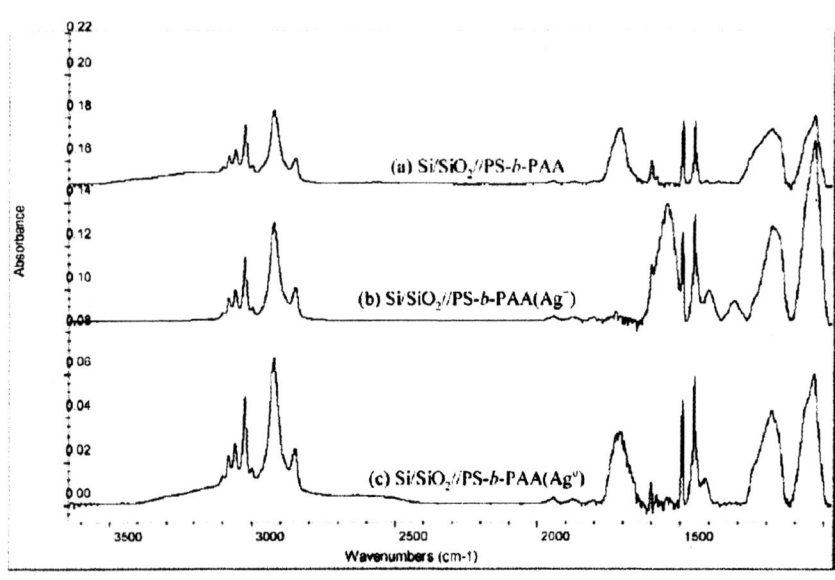

Figure 5. ATR-FTIR Spectra for the reduction of the Si/SiO$_2$//PS-b-PAA(Ag$^+$) polymer brush to produce silver nanoparticles.

XPS was used to confirm that the silver and palladium were still present in the diblock copolymer brushes after the reduction reaction and also to confirm reduction of the metal cations. The XPS spectra of the Si/SiO$_2$//PS-b-PAA(Ag$^+$) polymer brush after the reduction reaction had a silver content of 8.4 atomic%, compared with 9.6 atomic% before the reaction. The XPS spectra of the Si/SiO$_2$//PS-b-PAA(Pd^{2+}) polymer brush after reduction had a palladium content of 7.0 atomic%, compared with 6.6 atomic% before the reaction. These results indicate that metal content in the diblock copolymer brushes did not change significantly during the reduction. The small variation seen in metal contents can be explained by variations in sampling depth of the XPS and in the height of the PAA. Comparison of the XPS spectra for the Si/SiO$_2$//PS-b-PAA(Ag$^+$) polymer brush before and after reduction indicates a slight shift to lower binding energies for the silver 3d^3 and 3d^5 peaks. Similar results were seen for the palladium 3d^3 and 3d^5 peaks. These results suggest that the reduction conditions used do indeed result in a reduced form of the metals and, coupled with the ATR-FTIR results, provide strong evidence for the formation of zerovalent metal with in the diblock copolymer brushes.

Atomic force microscopy (AFM) was used to image the polyelectrolyte brushes both before and after the reduction reaction. Figure 6 depicts the AFM images for the Si/SiO$_2$//PS-b-PAA(Ag$^+$) polymer brush. Before the reduction reaction (Figure 6a), the AFM image indicates a relatively smooth surface with a root-mean-square (rms) roughness of 0.7 nm. After the reduction reaction (Figure 6b), the rms roughness increases to 3 nm and the AFM image clearly indicates the presence of defined surface features. The presence of prominent surface features and increase in the rms roughness are consistent with the formation of silver nanoparticles within the diblock copolymer brush. An analysis of many randomly drawn line profiles suggest an average height for the surface features in the reduced sample of 11 nm and for surface variations in the non-reduced sample of 4 nm. While this difference in height suggests particles that are 7 nm in diameter, we cannot confirm this as we do not know how the particles are embedded in the polymer brush. As the outer PAA layer is soft and the surface energy of Ag^0 is high, the particles could be at least partially embedded in the brush. To minimize interfacial energy, the particles are predicted to be spherical which is consistent with the 7 nm features in the AFM image after the reduction reaction. Similar results were seen for the reduction of the Si/SiO$_2$//PS-b-PAA(Pd^{2+}) polymer brush.

Summary

The use of ATRP to synthesize tethered polyelectrolyte brushes offers many advantages when compared to traditional radical polymerization methods. Using ATRP, the authors have been able to prepare both homopolymer and diblock copolymer polyelectrolyte brushes. Each of these systems demonstrated stimuli

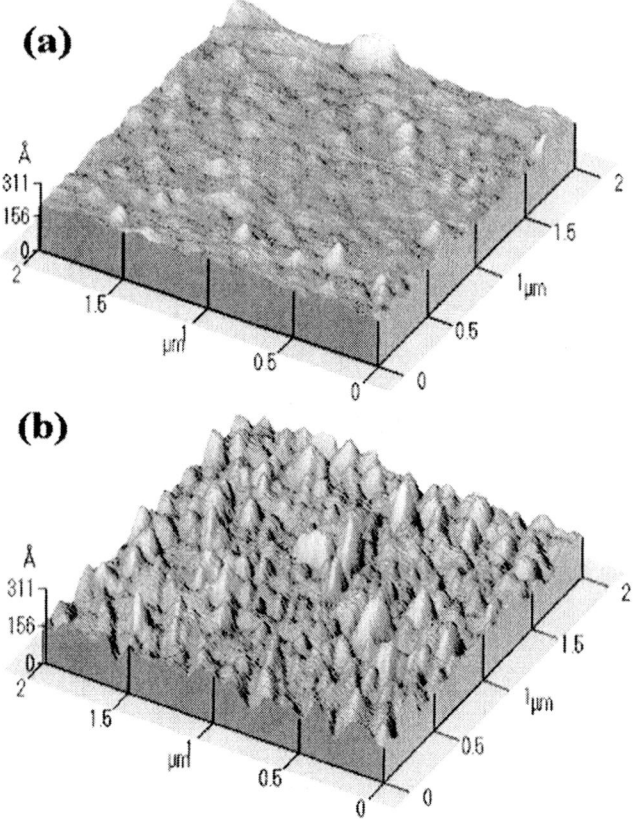

Figure 6. AFM images of the $Si/SiO_2//PS$-b-$PAA(Ag^+)$ polymer brush (a) before and (b) after reduction (See page 2 of color insert.)

responsive behavior in response to changes in either solvent or salt concentration. The diblock copolymer brushes were also used as templates for the preparation of both silver and palladium nanoparticles.

References

1. Zhulina, E. B.; Singh, C.; Balazs, A. C. *Macromolecules* **1996**, *29*, 6338-6348.
2. Zhao, B.; Brittain, W.J. *Prog. Polym. Sci.* **2000**, *25*, 677-710.

3. Edmondson, S.; Osborne, V. L.; Huck, W. T. S. *Chem. Soc. Rev.* **2004**, *33*, 14-22.
4. Rühe, J. In *Polymer Brushes*; Advincula, R. C.; Brittain, W. J.; Caster, K. C.; Rühe, J., Ed.; Wiley-VCH: Weinheim, Germany, 2004; pp 1-31.
5. Matyjaszewski, K.; Miller, P. J.; Shukla, N.; Immaraporn, B.; Gelman, A.; Luokala, B. B.; Siclovan, T. M.; Kickelbick, G.; Vallant, T.; Hoffmann, H.; Pakula, T. *Macromolecules* **1999**, *32*, 8716-8724.
6. Wu, T.; Efimenko, K.; Vlcek, P.; Subr, V.; Genzer, J. *Macromolecules* **2003**, *36*, 2448-2453.
7. Milner, S. T. *Science* **1991**, *251*, 905-914.
8. Halperin, A.; Tirrell, M.; Lodge, T. P. *Adv. Polym. Sci.* **1992**, *100*, 31-71.
9. Mansky, P.; Liu, Y.; Huang, E.; Russell, T. P.; Hawker, C. J. *Science* **1997**, *275*, 1458-1460.
10. Matyjaszewski, K. In *Handbook of Radical Polymerization*; Matyjaszewski, K.; Davis, T. P., Ed.; Wiley-Interscience: Hoboken, NJ, 2002; pp 361-406.
11. Boyes, S. G.; Brittain, W. J.; Weng, X.; Cheng, S. Z. D. *Macromolecules* **2002**, *35*, 4960-4967.
12. Kong, X.; Kawai, T.; Abe, J.; Iyoda, T. *Macromolecules* **2001**, *34*, 1837-1844.
13. Von Werne, T.; Patten, T. E. *J. Am. Chem Soc.* **2001**, *123*, 7497-7505.
14. Osborne, V. L.; Jones, D. M.; Huck, W. T. S. *Chem. Commun.* **2002**, 1838-1839.
15. Chen, X.; Randall, D. P.; Perruchot, C.; Watts, J. F.; Patten, T. P.; von Werne, T.; Armes, S. P. *J. Coll. Inter. Sci.* **2003**, *257*, 56-64.
16. Biesalski, M.; Rühe, J. *Macromolecules* **2004**, *37*, 2196-2202.
17. Boyes, S. G.; Akgun, B.; Brittain, W. J.; Foster, M. D. *Macromolecules* **2003**, *36*, 9539-9548.
18. Wu, T.; Genzer, J.; Gong, P.; Szleifer, I.; Vlček, P.; Šubr, V. In *Polymer Brushes*; Advincula, R. C.; Brittain, W. J.; Caster, K. C.; Rühe, J., Ed.; Wiley-VCH: Weinheim, Germany, 2004; pp 287-315.
19. Zhao, B.; Brittain, W. J. *J. Am. Chem. Soc.* **2000**, *122*, 2407-2408.
20. Zhao, B.; Brittain, W. J. *Macromolecules* **2000**, *33*, 8821-8827.
21. Mayer, A. B. R.; Mark, J. E. *Colloid Polym. Sci.* **1997**, *275*, 333-340.
22. Schmidt, H. *Appl. Organometal. Chem.* **2001**, *15*, 331-343.
23. Joly, S.; Kane, R.; Radzilowski, L.; Wang, T.; Wu, A.; Cohen, R. E.; Thomas, E. L.; Rubner, M. F. *Langmuir* **2000**, *16*, 1354-1359.
24. Zong, C.-J.; Maye, M. M. *Adv. Mater.* **2001**, *13*, 1507-1511.
25. Wang, T. C.; Rubner, M. F.; Cohen, R. E. *Langmuir* **2002**, *18*, 3370-3375.

Chapter 5

Gradient Stimuli-Responsive Polymer Grafted Layers

Sergiy Minko[1], Leonid Ionov[2], Alexander Sydorenko[2], Nikolay Houbenov[2], Manfred Stamm[2], Bogdan Zdyrko[3], Viktor Klep[3], and Igor Luzinov[3]

[1]Chemistry Department, Clarkson University, 8 Clarkson Avenue, Potsdam, NY 13699
[2]Institute for Polymer Research, Hohe Strasse 6, 01069 Dresden, Germany
School of Materials Science and Engineering, 263 Sirrine Hall, Clemson University, Clemson, SC 29634–0971

We discuss our recent results on fabrication and study of responsive thin polymer films covalently grafted to solid substrates. The films are prepared from two different incompatible polymers (mixed polymer brush), where the film composition gradually changes along a one direction of the sample surface. The gradual change of the composition causes the gradual change of wetting behavior. At the same time incompatibility of the polymers introduces the switching behavior due to the phase segregation mechanisms. The switching is sensitive to external signals (solvent, temperature, pH). Overlap of these two phenomena allows us to design surfaces with switchable gradients of wetting.

Introduction

Surfaces responsive to external signals (light, electric field, solvent, pH [1-6]), attract high interest for regulation and switching of wettability, adhesion, adsorption, etc.[7-9] Very important example of stimuli responsive surfaces is represented by mixed polymer brushes consisting of two or more incompatible polymers grafted to the same substrate. The successful synthesis of mixed polymer brushes via "grafting to" and "grafting from" approaches was recently reported.[10-12] Theoretical and experimental studies give evidence that the mixed brushes undergo phase transition upon external stimuli. The properties of the thin film depend on the balance between layered and lateral phase segregation mechanisms.[13-16] Selective solvents (good solvent for one of two grafted polymers) stabilize a dimple like morphology: unfavorite polymer forms clusters segregated to the grafting surface while the favorite polymer forms a discontinuous phase preferentially segregated to the top. In non-selective solvents the lateral segregation dominates and the mixed brushes form a ripple-like morphology: lamellar like stripes formed by alternating microphases of the both polymers where both types of macromolecular chains are exposed on the surface. The solvent induced phase transition leads to switching of surface properties of polymer brushes between that of two different polymers, for example hydrophilic and hydrophobic[7,10,17], soft and hard [18,19], sticky and non sticky.[7]

Here we overview a recent extension made in the field of mixed polymer brushes which addresses new possibilities for applications and investigation of the switching mechanism. We discuss the mixed brushes where the composition gradually changes along the sample ("gradient brushes", GB). GB were suggested as a combinatorial approach [20,21] to study of brush properties as a function of grafting density and molecular weight. Gradient mixed brushes attract additional interest because of the possibility to switch the gradient. Design of polymer grafted layers with gradually changing switching properties will allow construction of miniaturized devices and smart materials sensitive to external signals.

Synthesis of "Gradient Brushes"

Carboxyl terminated polystyrene (PS), poly(*tert*-butyl acrylate) (PBA), poly(2-vinylpyridine) (P2VP) purchased from Polymer Source were used for the fabrication of the polymer brushes by "grafting to" method. The detailed description of used materials and synthetic procedures is given elsewhere [7,11,22-30].

Preparation of polystyrene gradient brush. Our synthetic procedure includes the first step of the deposition of either ω-functional alkysilane 3-

glycidoxypropyltrimethoxysilane (GPS) or polymer anchoring layer on the surface of a solid substrate. The anchoring layer (2-3 nm thick) is made from poly(glycidyl methacrylate) (PGMA). The second step is the grafting of the carboxyl terminated PS by esterification reaction at temperature that is above the glass transition temperature (T_g) of PS. PS film is deposited on the top of PGMA layer and heated above T_g[26,30]. The gradient of grafting density is introduced employing gradient of temperature on a specially designed stage [25,27]. The temperature gradient stage was constructed from a stainless steel plate where two opposite sides of the plate were connected to the heating and cooling elements (**Figure 1a**) [31-34]. The temperature of the stage is monitored by five pairs of thermocouples located along the sample. The stage is equipped with a cover so that the sample is under N2 atmosphere. The gradient stage provides a linear temperature increase from the cool end to the hot end (**Figure 1b**). The temperature gradient affects the temperature dependent kinetics of grafting. Thus the grafting amount gradually changes from the cool end to the hot end.

Preparation of mixed gradient brushes. Polyacrylic acid (PAA) - P2VP (PAA-mix-P2VP) and PS -P2VP (PS-mix-P2VP) brushes were synthesized using an additional grafting step. Initially, PBA and PS gradient brushes were grafted, respectively. Afterward, P2VP was grafted in the second step. The carboxyl terminated P2VP was deposited on the top of the gradient PS brush and heated up above T_g. Once the gradient PBA-mix-P2VP brush was fabricated PBA was hydrolyzed yielding the PAA-mix-P2VP gradient brush [17,25-27,30,35].

Sample Characterization. Thickness of the PGMA, PS, PBA layers and the mixed brushes was measured at λ=633 nm and an angle of incidence of 70° with a SENTECH SE - 402 microfocus ellipsometer (lateral resolution is defined by the beam spot of about 20 μm)[17,23,27,28,30,35]. AFM studies were performed with a Dimension 3100 (Digital Instruments, Inc., Santa Barbara, CA) microscope. The tapping and phase modes were used to map the film morphology at the ambient conditions. Advancing contact angles of water were measured using a "DSA – 10" Krüss equipment. Since typical layers obtained in this work exhibited very small roughness (measured by AFM) ranging from 0.3 nm to 3 nm, the Cassie equation was applied to estimate the surface composition [36]:

$$\cos\theta = \varphi \cos\theta_A + (1-\varphi) \cos\theta_B, \tag{1}$$

where θ is the water contact angle, θ_A and θ_B are the reference contact angles of water on surfaces of the type A and the type B, respectively, φ is the fraction of the surface covered by A. An advancing contact angle of water on PGMA was 66°.

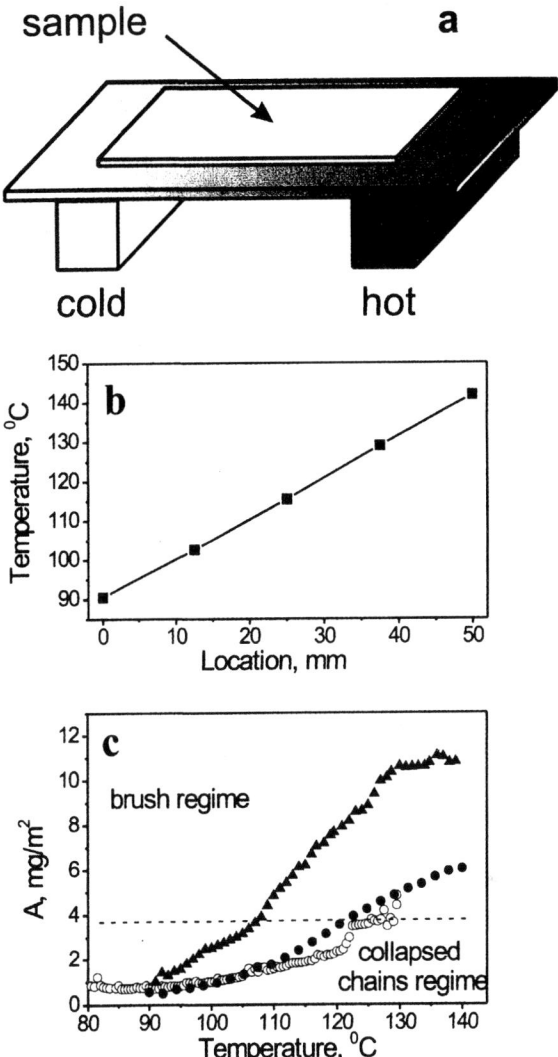

Figure 1. Principal scheme of temperature gradient stage (a) and a temperature gradient formed on the surface of the stage; (b) examples of PS brushes: the curves are presented in terms of grafted amount vs. grafting temperature on the stage.[27] (Reproduced with permission from reference 27. Copyright 2004.)

Monocomponent Gradient Grafted Polymer Layers

There are several major parameters, which control the grafting density of polymer chains: temperature, grafting time, molecular weight of the polymer, and surface density of anchoring sites on the substrate. In our experiment we keep unchanged grafting time, density of anchoring sites, and molecular weight of the polymer, while the grafting temperature is changing gradually along the substrate and, thus, is a commanding parameter to design the gradient of grafting density.

We applied the temperature gradient from 90°C to 140°C for grafting PS (**Figure 1b**). Carboxyl terminated PS effectively react with the PGMA modified substrate at temperatures above the glass transition temperature ($T_{g,\ PS}$ bulk =97°C- 100°C[37]) due to the significant enhancement in mobility of the chains. The temperature dependence of the grafting kinetics results in a gradual increase of the grafting amount (A). The grafting levels off at A values usually not larger than 10 mg/m.^{2} In **Figure 1c** we present the results of the ellipsometric mapping of the PS grafted layers attached to the PGMA and GPS anchoring layers. In fact, we detected only very small grafting ($A \approx 1$ mg/m^2) at 60-95°C. The significant increase in the grafted amount was obtained at T_g and the saturation level, A_{max}, was approached in 8 h of the grafting. The A_{max} value depended on the nature of the anchoring layer. In the case of GPS, A_{max} was about 4.5 mg/m^2, which is in good agreement with data reported elsewhere.[23] A_{max} in the case of the PGMA anchoring layer was 2-3 times larger. This experimental fact was recently studied in detail by Iyer et al.[30] In brief, the grafting on PGMA layer results in the formation of a fractal interface where the grafted points are located at different levels into the PGMA primary film. In contrast to SAM of GPS, PGMA epoxy groups are distributed in the layer over the entire thickness and obviously become more available for PS-COOH end reactive groups. It effectively leads to higher density of accessible anchoring sites.

The gradient of grafting density introduces a gradual change of the thin film properties. One very important example is a gradual change of surface energy and wetting behavior. At a low grafting density water penetrates the layer and contacts the substrate. A larger grafting density results in a less contact with the substrate. Finally, at high grafting densities water contacts only PS.

The gradient of wetting behavior of the samples was investigated using the contact angle (CA) measurements in different locations over the surface of the grafted PS layers. The resolution of this method is limited by the size of water drops, however, it gives a general trend for the variation of the surface energy along the substrate modified with the gradient films.

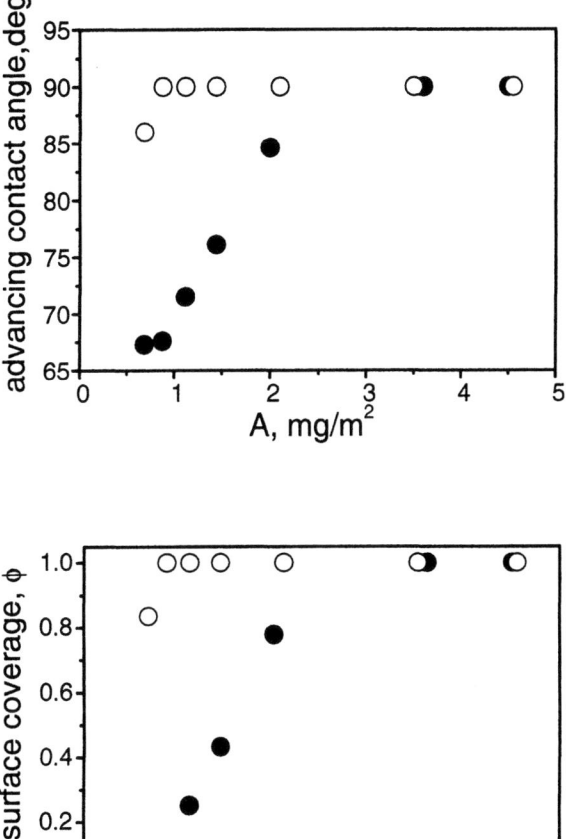

Figure 2. Advancing contact angle vs. location along the substrate (in terms of grafting amount in the location of the probing water drop) modified with the grafted PS-COOH layers (top). Surface coverage (calculated using equation 1) vs. grafting amount of PS (bottom). Before measurements the layers were exposed to toluene (open circles), ethanol (solid circles), and dried.

We found out that wetting behavior of the gradient film depended on the history of the sample preparation, namely on solvent which was used to rinse the gradient films. A gradual change of CA values on the samples modified with the PS layers was observed after treatment of the samples with poor solvent (ethanol) (**Figure 2**). At A of about 3 mg/m^2 the slope decreases and the contact angle value approaches the plateau.[38] This plateau grafting density corresponds to the formation of the dense grafted layer. The contact angle value 85° at this point is smaller as compared to the value for a reference PS film (90°). Further increase of grafting density results in a slow increase of contact angle until it approaches the value of 90°. However, no signs of a wetting gradient were observed upon treatment of the sample with good solvent (toluene). The transition is reversible. The gradient appears again upon exposure to ethanol.

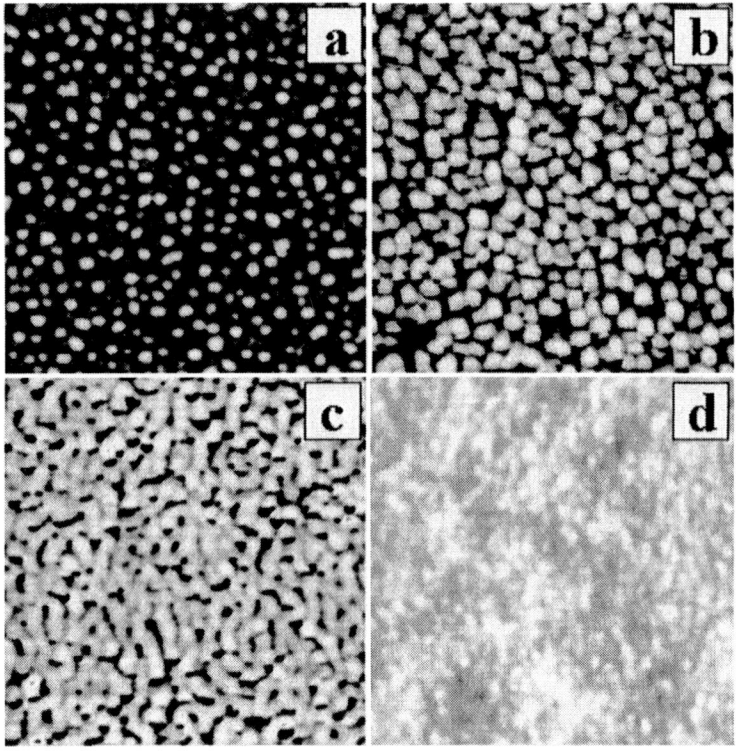

Figure 3. AFM images (500 x 500 nm) of PS grafted layers of different grafting densities. a – 0.6·mg/m^2, b – 1,8·mg/m^2, c – 2.7·mg/m^2, d – 4.7 mg/m^2. Anchoring layer – PGMA. Samples were treated with ethanol. Vertical scale is 15 nm. (Reproduced with permission from reference 27. Copyright 2004.)

This result clearly demonstrates the switching behavior of the monocomponent grafted layer at intermediate grafting densities. The PS films were smooth and homogeneous after exposure to a good solvent (toluene) over the entire investigated region. The values of rout-mean-square roughness (RMS) measured with AFM were in the range of 0.3-0.5 nm on the 1 μm² sample area. Meanwhile, AFM images of the samples revealed a pinned micelle morphology at low grafting densities (**Figures 3a** and **3b**). The lateral size of the micelles was in the range of 12-20 nm. The average micelle height was about 5.4±1.5 nm. Assuming a disc shape geometry of the micelles and that the micelle density is equal to the bulk density of PS, the averaged number of chains N_c in the micelles was evaluated to be 10:

$$N_c = (\pi R^2 h \rho) / (M_w/N_A) \qquad (2)$$

The increase of the surface coverage resulted in the transformation of the micelles into the homogeneous brush-like layer (**Figure 3c and 3d**) at the grafting densities larger than 3-4 mg/m². Thus, we conclude that the origin of the wettability switching at intermediate grafting densities is in the transition between morphologies of pinned micelles and the homogenous brush-like layer.

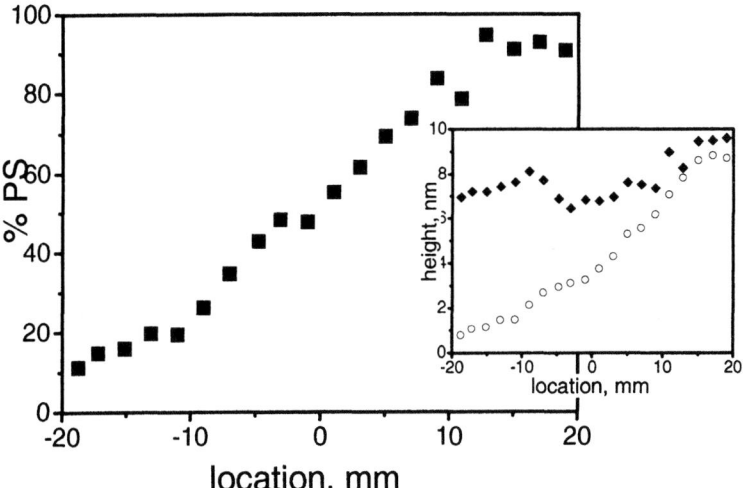

Figure 4. Fraction of grafted polystyrene vs. location of the measured spot (on X axis) for the gradient PS-mix-P2VP brush. The inset demonstrates the ellipsometric thickness of the homopolymer PS (open circles) and PS-mix-P2VP(solid squares) brushes.

Mixed Gradient Brushes

The described above synthetic approach was successfully used for the synthesis of mixed GB. We discuss here two different examples of the gradient mixed brushes.

The first example is represented by the mixed brush prepared from PS and P2VP (PS-*mix*-P2VP). The result of the ellipsometric mapping of the 1D gradient PS-*mix*-P2VP is shown in **Figure 4**. The thicknesses of the gradient PS brush (synthesized in the first grafting step) and PS-*mix*-P2VP (synthesized on the second grafting step) are presented in the inset. These two plots in the inset were used to calculate the composition of the gradient PS-*mix*-P2VP brush.

The thickness of the PS layer gradually increases along the X-axis from the left-hand side to the right-hand side reflecting the change of grafting density of PS. The inset demonstrates that the entire thickness of the mixed brush along the X-axis is almost constant value[30] of about 7 nm, corresponding to the grafting density of approximately 0.1 chains/nm^2. The distance between grafting sites equal to 3.5 nm is smaller than the gyration radius (Rg) of PS and P2VP polymer coils ($Rg \sim 5$ nm, θ-conditions). Consequently, the grafted polymer film can be considered as a brush-like layer.

The gradual change of the mixed brush composition was confirmed by the contact angle measurements performed on the dry brush surface upon exposure to chloroform (nonselective solvent) for 10 min [17,39]. The water contact angle increases almost linearly with the increase of the PS fraction (**Figure 5a**). The composition of the top of the brush calculated using the Cassie equation is very close to the composition obtained from the ellipsometric data (**Figure 5b**).

However, the treatment of the brush with selective solvents (toluene and ethanol) reveals the unique switching behavior of the mixed brush. The gradient can be switched reversibly from the case when the wetting gradually increases from the left-hand side to the right hand side to the case when the wetting is almost constant along the X-axis (**Figure 5**). In other words, we can effectively "turn on" and "turn off" the wetting gradient upon treatment with the solvent (**Figure 5b**). This switching is limited by the brush composition and can be obtained within a reasonable compositional asymmetry of the mixed brush (from 30% to 70% of PS). The mechanism of the switching effect is explained in **Figure 6**. In selective solvent the favorite polymer is preferentially on the top at different compositions. The gradient is "turned off". In non-selective solvent both polymer are on the top and the composition of the GB gradually changes along X axis. The gradient is "turned on".

The second example is represented by the mixed polyelectrolyte (PE) PAA-*mix*-P2VP brush which demonstrates a unique switching behavior in aqueous environment at different pH (**Figure 7**). The mixed PE brush can be either hydrophilic or hydrophobic depending on pH and composition. The PAA-*mix*-

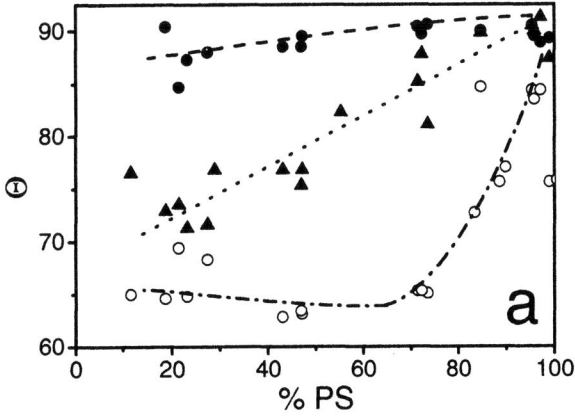

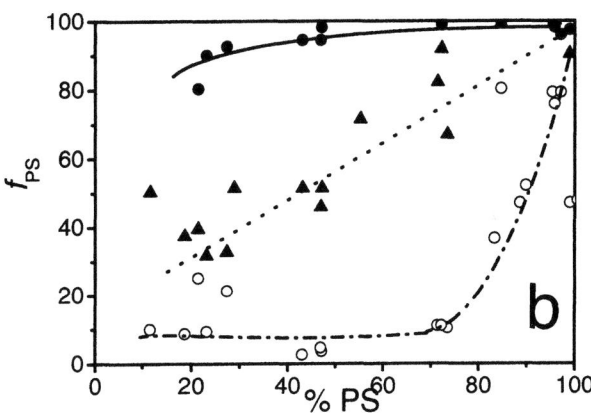

Figure 5. Switching of water contact angle (Θ) of the gradient PS-mix-P2VP brush after exposure to different solvents (a) and the fraction of PS (f_{PS}) on the top of the brush estimated with Cassie equation (solid circles- toluene; solid triangles- chloroform; open circles - ethanol) (b) vs. the measurement location along X axis expressed in terms of PS fraction.

P2VP brushes of compositions (ranging from 20% to 60% of PAA) close to the symmetrical case were highly hydrophilic upon treatment with both low and high pH water, however they were hydrophobic upon treatment with neutral water (see **Figure 7 b,c,d**). If the brush is strongly asymmetric, the switching behavior is dominated by a major component. For example, if P2VP is a major component the mixed brush is hydrophilic at low pH, however it is hydrophobic at high pH (**Figure 7a**). If PAA is a major component the inverse wetting behavior was observed (**Figure 7e,f**). The contact angle value increases at low and high pH with the increase of fraction of PAA and P2VP, respectively. Therefore, the switching range (difference between water advancing contact angles obtained upon treatment with water at low and high pH aqueous solutions) depends on the brush composition.

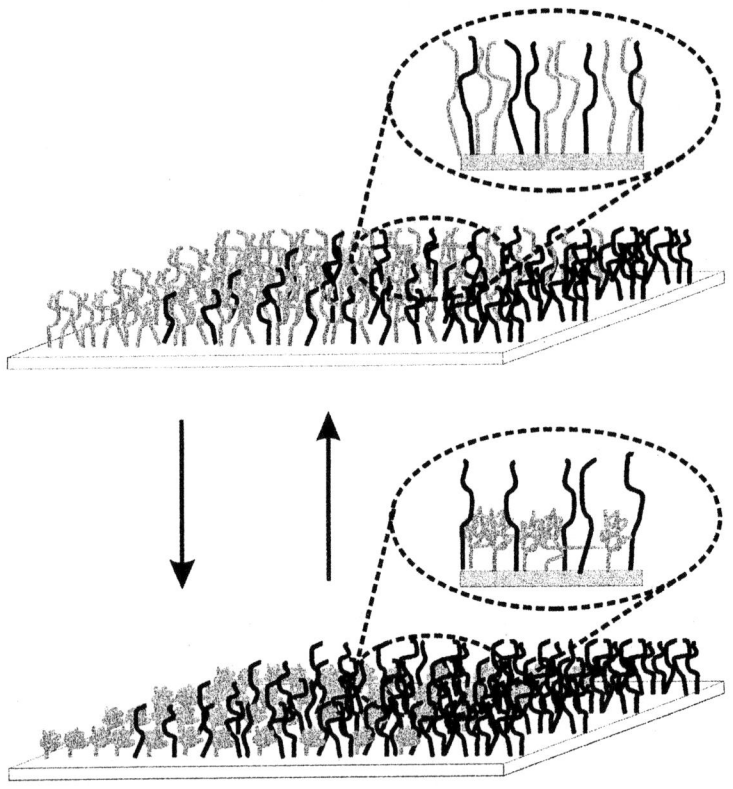

Figure 6. Scheme of the gradient mixed brush morphology upon treatment with nonselective (top) and selective (bottom) solvents. In nonselective solvents both polymers are on top: the gradient is "turned on." In selective solvent only a favorite polymer is on top: the gradient is "turned off".

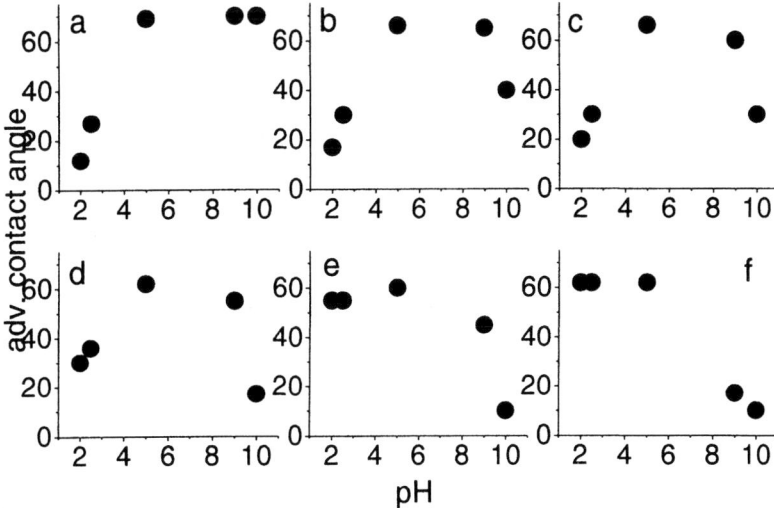

Figure 7. Switching of water contact angle of the gradient PAA-mix-PVP brush at the locations with different fraction of PAA (a – 20%, b – 30%, c – 40%, d – 60%, e – 80%, f – 100%) vs. pH.

The dependence of the switching behavior on the brush composition is a fine instrument for manipulations with the wetting gradient. In fact, the wetting gradient could be switched as shown in **Figure 8**. At pH 2 water advancing contact angle gradually increases from the left hand side of the sample to the right hand side. An inverse situation was obtained at pH 10. In the latter case water advancing contact angle gradually decreases from the left hand side to the right hand side of the sample. In other words, the direction of the gradient can be altered using a pH signal. The absolute value of the wetting gradient (difference between water advancing contact angle on the left hand side and on the right hand side of the sample) can be tuned by a small variation of pH as can be seen from **Figure 8**.

On the basis of the previous reports,[13,39-42] the switching effect is explained by the change of morphology and composition of the top layer of the brush. The mechanism of the switching behavior of the mixed PE brush is illustrated in **Figure 9** [35]. Each homopolymer in the mixed brush is a weak PE. The charge density of the weak PE depends on pH. P2VP is protonated at low pH. The degree of protonation and the charge density of P2VP chains increase with a decrease of pH. A similar scenario can be applied to PAA chains, but PAA is dissociated at high pH. Both polymers form a polyelectrolyte complex in the pH range from 5 to 6. At low pH, PAA chains adopt a compact conformation on the bottom of the film while P2VP chains are highly protonated and stretched away

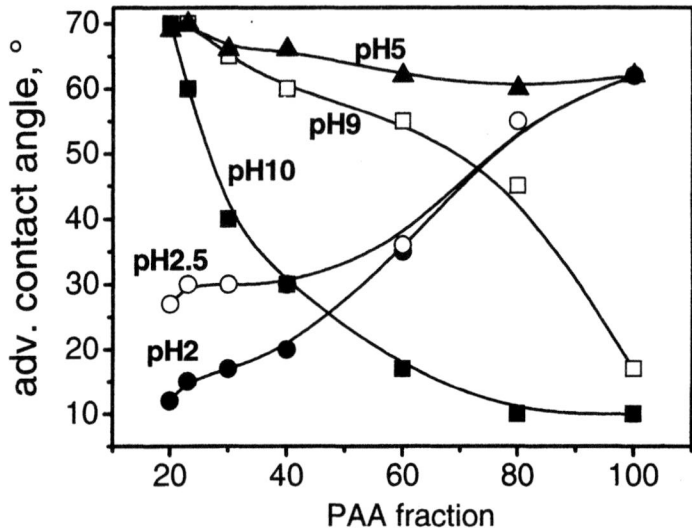

Figure 8. Switching of water contact angle of the gradient PAA-mix-PVP brush vs. composition upon exposure to water at different pH.

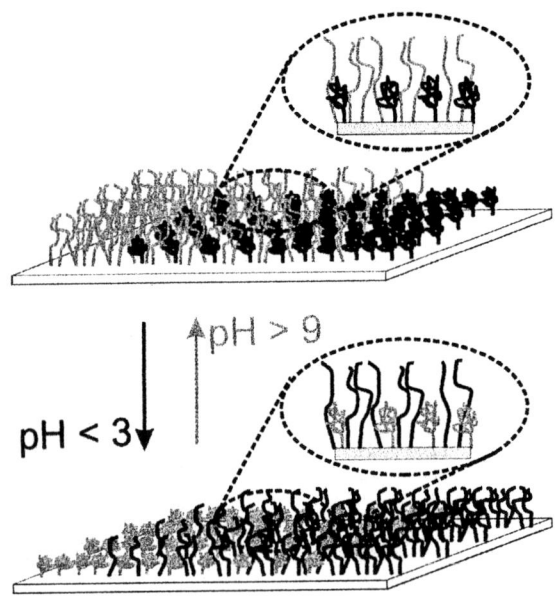

Figure 9. Schematic representation of switching behavior of mixed PE brush upon change of pH: low pH (a) and high pH (b).

from the surface. They preferentially occupy the top of the brush. At high pH values, the inverse transformation takes place. P2VP is collapsed on the bottom, while negatively charged PAA chains are on the top. Consequently, the change of pH affects the change of the charge density gradient. The latter introduces the wetting gradient.

In summary, we developed the novel approach for synthesis of mono component and mixed polymer brushes with a unidirectional gradual change of grafting density and composition, respectively. This approach allows for a fast screening of the switching behavior of mixed brushes. The reported switching behavior of mono- and bi- component polymer grafted layers is of potential interest for a combinatorial analysis of interactions in aqueous environment, in biological systems, responsive microdevices, and drug delivery systems.

References

1. Russell, T. P. In *Science*, 2002; Vol. 297, pp 964-967.
2. Ito, Y.; Ochiai, Y.; Park, Y. S.; Imanishi, Y. In *Journal of the American Chemical Society*, 1997; Vol. 119, pp 1619-1623.
3. Ionov, L.; Minko, S.; Stamm, M.; Gohy, J. F.; Jerome, R.; Scholl, A. In *Journal of the American Chemical Society*, 2003; Vol. 125, pp 8302-8306.
4. Nath, N.; Chilkoti, A. In *Adv. Mat.*, 2002; Vol. 14, pp 1243-1246.
5. Anastasiadis, S. H.; Retsos, H.; Pispas, S.; Hadjichristidis, N.; Neophytides, S. In *Macromolecules*, 2003; Vol. 36, pp 1994-1999.
6. Ichimura, K.; Oh, S.; Nakagawa, M. In *Science*, 2000; Vol. 288, p 1624.
7. Minko, S.; Muller, M.; Motornov, M.; Nitschke, M.; Grundke, K.; Stamm, M. In *Journal of the American Chemical Society*, 2003; Vol. 125, pp 3896-3900.
8. Matthews, J. R.; Tuncel, D.; Jacobs, R. M. J.; Bain, C. D.; Anderson, H. L. In *J. Am. Chem. Soc.*, 2003; Vol. 125, pp 6428-6433.
9. Feng, C. L.; Zhang, Y. J.; Jin, J.; Song, Y. L.; Xie, L. Y.; Qu, G. R.; Jiang, L.; Zhu, D. B. In *Langmuir*, 2001; Vol. 17, pp 4593-4597.
10. Sidorenko, A.; Minko, S.; Schenk-Meuser, K.; Duschner, H.; Stamm, M. In *Langmuir*, 1999; Vol. 15, pp 8349-8355.
11. Minko, S.; Patil, S.; Datsyuk, V.; Simon, F.; Eichhorn, K. J.; Motornov, M.; Usov, D.; Tokarev, I.; Stamm, M. In *Langmuir*, 2002; Vol. 18, pp 289-296.
12. Zhao, B. In *Polymer*, 2003; Vol. 44, pp 4079-4083.
13. Minko, S.; Muller, M.; Usov, D.; Scholl, A.; Froeck, C.; Stamm, M. In *Physical Review Letters*, 2002; Vol. 88, pp -.
14. Marko, J. F.; Witten, T. A. In *Phys. Rev. Lett.*, 1991; Vol. 66, pp 1541-1544.

15. Zhulina, E.; Balazs, A. C. In *Macromolecules*, 1996; Vol. 29, pp 2667-2673.
16. Muller, M. In *Physical Review E*, 2002; Vol. 65, pp -.
17. Minko, S.; Patil, S.; Datsyuk, V.; Simon, F.; Eichhorn, K. J.; Motornov, M.; Usov, D.; Tokarev, I.; Stamm, M. *Langmuir* **2002**, *18*, 289-296.
18. Lemieux, M.; Minko, S.; Usov, D.; Stamm, M.; Tsukruk, V. V. *Langmuir* **2003**, *19*, 6126-6134.
19. Lemieux, M.; Usov, D.; Minko, S.; Stamm, M.; Shulha, H.; Tsukruk, V. V. *Macromolecules* **2003**, *36*, 7244-7255.
20. Tomlinson, M. R.; Genzer, J. *Chem. Commun.* **2003**, 1350-1351.
21. Wu, T.; Efimenko, K.; Genzer, J. *J. Am. Chem. Soc.* **2002**, *124*, 9394-9395.
22. Ionov, L.; Houbenov, N.; Sidorenko, A.; Luzinov, I.; Minko, S.; Stamm, M. *Langmuir, accepted*.
23. Luzinov, I.; Julthongpiput, D.; Malz, H.; Pionteck, J.; Tsukruk, V. V. *Macromolecules* **2000**, *33*, 1043-1048.
24. Minko, S.; Luzinov, I.; Patil, S.; Datsyuk, V.; Stamm, M. *Abstr Pap Am Chem S* **2001**, *222*, U370-U370.
25. Zdyrko, B.; Klep, V.; Luzinov, I.; Minko, S.; Sydorenko, A.; Ionov, L.; Stamm, M. *Polymer Preprints* **2003**, *44*, 522-523.
26. Zdyrko, B.; Klep, V.; Luzinov, I. *Langmuir* **2003**, *19*, 10179-10187.
27. Ionov, L.; Zdyrko, B.; Sidorenko, A.; Minko, S.; Klep, V.; Luzinov, I. *Macromol. Rapid Comm.* **2004**, *25*, 360-365.
28. Ionov, L.; Sidorenko, A.; Stamm, M.; Minko, S.; Zdyrko, B.; Klep, V.; Luzinov, I. *Macromolecules* **2004**, *ASAP article*.
29. Sidorenko, A.; Minko, S.; Schenk-Meuser, K.; Duschner, H.; Stamm, M. *Langmuir* **1999**, *15*, 8349-8355.
30. Iyer, K. S.; Zdyrko, B.; Malz, H.; Pionteck, J.; Luzinov, I. *Macromolecules* **2003**, *36*, 6519-6526.
31. Smith, A. P.; Douglas, J. F.; Meredith, J. C.; Amis, E. J.; Karim, A. *J. Polym. Sci Pol. Phys.* **2001**, *18*, 2141-2158.
32. Meredith, J. C.; Smith, A. P.; Karim, A.; Amis, E. J. *Macromolecules* **2000**, *33*, 9747-9756.
33. Meredith, J. C.; Karim, A.; Amis, E. J. *Macromolecules* **2000**, *33*, 5760-5762.
34. Smith, A. P.; Meredith, J. C.; Douglas, J. F.; Amis, E. J.; Karim, A. *Phys Rev Lett* **2001**, *87*, 015503.
35. Houbenov, N.; Minko, S.; Stamm, M. *Macromolecules* **2003**, *36*, 5897-5901.
36. Cassie, A. *Discuss. Faraday Soc.*. **1948**, *3*, 11.
37. Bliznyuk, V. N.; Assender, H. E.; Briggs, G. A. D. *Macromolecules* **2002**, *35*, 6613-6622.
38. Milner, S. T. *Science* **1991**, *251*.
39. Sidorenko, A.; Minko, S.; Schenk-Meuser, K.; Duschner, H.; Stamm, M. In *Langmuir*, 1999; Vol. 15, pp 8349-8355.

40. Minko, S.; Patil, S.; Datsyuk, V.; Simon, F.; Eichhorn, K. J.; Motornov, M.; Usov, D.; Tokarev, I.; Stamm, M. In *Langmuir*, 2002; Vol. 18, pp 289-296.
41. Minko, S.; Usov, D.; Goreshnik, E.; Stamm, M. In *Macromolecular Rapid Communications*, 2001; Vol. 22, pp 206-211.
42. Lemieux, M.; Usov, D.; Minko, S.; Stamm, M.; Shulha, H.; Tsukruk, V. V. In *Macromolecules*, 2003; Vol. 36, pp 7244-7255.

Chapter 6

Tailoring of Thin Polymer Films Chemisorbed onto Conductive Surfaces by Electrografting

C. Jérôme and R. Jérôme

Center for Education and Research on Macromolecules, University of Liege, B6 Sart-Tilman, B–4000 Liege, Belgium

Cathodic electrografting is an efficient technique to impart adhesion to poly(meth)acrylate coatings onto inorganic conducting surfaces. Although this technique was restricted for many years to very few monomers ((meth)acrylonitrile) and to deposition of very thin polymer films, recent developments have overcome these limitations. First of all, the judicious choice of the solvent has proved to be a powerful lever to increase the range of the chemisorbed polymers, including functional polymers. Quite interestingly, classical controlled polymerization techniques have b een c ombined w ith c athodic electrografting as a powerful strategy for tuning thickness, properties and reactivity of the chemisorbed organic films. At the time being, cathodic electrografting has contributed to substantial progress in very demanding applications, such as protection against corrosion, food packaging, biomaterials, sizing of reinforcing agents in polymer composites and sensoring devices.

1. Fundamentals of Cathodic Electrografting

1.1 Experimental Facts

Fifteen years ago, the possible grafting of polyacrylonitrile (PAN) in acetonitrile onto a common metal (Ni) was reported for the very first time as results of the careful control of the cathodic polarization [1]. This breakthrough aroused not only curiosity but also optimistic prospects for improving the polymer coating of inorganic substrates.

Because acetonitrile is a non solvent for PAN, the original process has been carried out in a good solvent for it, i.e., dimethylformamide (DMF). Two distinct reduction phenomena are observed (Figure 1, unbroken line) [2]. At the less cathodic peak, a polyacrylonitrile film is chemisorbed onto the cathode, whereas at the second reduction wave, the PAN chains grow in solution, and the original electrode surface is restored as result of the degrafting of the previously chemisorbed chains (see pictures in Figure 1). These observations have been confirmed by repeating the same experiment with a vibrating quartz crystal cathode (Figure 1, dotted line). Indeed, the vibration frequency of this cathode decreases at the potential of the first peak, in accordance with the PAN grafting, and the initial frequency is restored at the potential of the second reduction wave [3].

Thickness of PAN films deposited in the potential range of peak I, has been measured by ellipsometry [4], and found to be smaller in ACN (less than 25nm) than in DMF. In this solvent, the thickness increases from 25 to 150nm when the monomer concentration is increased from 0.1 to 2 M. In acetonitrile, the growing PAN chains precipitate on the electrode as soon as their length exceeds a critical value and they do not propagate anymore whatever the monomer concentration. The situation is completely different in DMF, a good solvent for the growing chains, that remain solvated during polymerization and can reach higher molecular weight at higher monomer concentration. Consistently, the morphological features of the cathode surface [5] can be seen by Atomic Force Microscopy (AFM) beneath the film deposited in acetonitrile, which is no longer the case when the film is formed in DMF at high monomer concentration.

The key characteristic feature of this process is thus the strong adhesion between a thin insulating polymer film deposited at the surface of a conducting substrate polarized at an appropriate potential [6]. The organic film is not only formed in a good solvent for the polymer but it resists peeling-tests, an adhesive rupture being systematically observed between the polymer and the adhering tape used for the testing and never at the polymer-substrate interface.

1.2 Electrografting Mechanism

The mechanism of this very unusual electrochemical reaction has been a matter of controversy. Lécayon et al. [7] proposed a mechanism that consists of the transfer of one electron from the cathode to the monomer, with the bonding of the radical-anion species to the metal. Chain initiation is thus an electrochemical event, in contrast to chain propagation which proceeds through

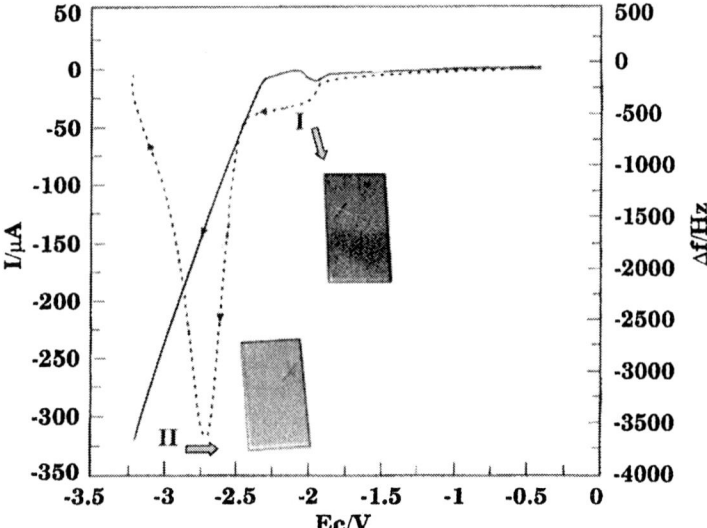

Figure 1. Potential dependence of the AN reduction current (unbroken line) and the quartz crystal frequency (dotted line) in DMF containing tetraethylammonium perchlorate (0.05M) and AN (0.2M). The potential scanning rate was 20mV/s. Insert: steel surface when the scanning was stopped at the potential of peak I (PAN chemisorbed on the area dipped in solution) and peak II (degrafting of PAN), respectively.

the repeated addition of the monomer to the chemisorbed anionic species [Ni-CH$_2$CH(CN)$^-$]. According to Jérôme et al. [4,8], a radical species would be grafted to the cathode surface and would propagate the AN polymerization. A series of experiments have been devised and carried out in order to discriminate between anionic and radical species [2,9]. The main conclusions can be summarized as follows: (i) the addition of a radical scavenger (deep-violet diphenylpicrylhydrazyl, DPPH) does not result in the end-capping of the PAN chains growing at the potential of peak I, (ii) no copolymerization of AN is observed at peak I neither with ε-caprolactone (anionic copolymerization) nor with vinyl acetate (radical copolymerization), which may merely indicate that none of the two comonomers is incorporated in the electrical double layer of the cathode, (iii) the addition of a transfer agent (CDCl$_3$) leads to the end-capping of the PAN chains by -D and not with -CCl$_3$, in agreement with an anionic propagation, (iv) when methyl methacrylate (MMA) is substituted for AN, part of the PMMA chains are growing in solution at the potential of peak I. These chains have been characterized by thermal gravimetric analysis because the TGA profile depends on the polymerization mechanism (anionic vs. radical). The one-step degradation profile which is actually observed, is typical of anionically prepared PMMA. Additionally, careful electrochemical studies were performed by using electrochemical quartz crystal microbalance that allow better understanding of the reaction processes [10-11].

All in all, the experimental observations are consistent with scheme 1, thus with the cathodic reduction of the acrylic monomer into a chemisorbed anion onto the cathode as long as the potential is in the range of peak I. In other words, an equilibrium between chemisorbed radical-anions and unbound radical anions is set-up, which depends very closely on the cathodic potential. The very low intensity of peak I is the signature of the electrodeposition of an insulating polymer. At a more cathodic potential, the chemisorption is no longer effective, the intensity of the reduction peak II increases rapidly, and the fast coupling of

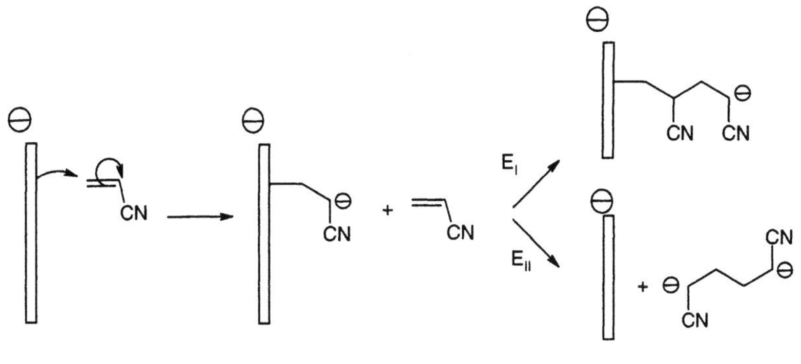

Scheme 1: Schematic mechanism for the electrografting of AN

the radical-anions into anionic dimers is followed by chain propagation in solution. Although this general mechanism accounts for the major experimental observations, additional experiments are needed to confirm it and to have a deeper insight on it.

1.3 The Origin of the Adhesion

Polyacrylonitrile films and the PAN/metal interface have been analyzed by powerful spectroscopic techniques, i.e., XPS, IR, ellipsometry, ultraviolet photoemission spectroscopy (UPS), and electron induced X-Ray emission spectroscopy (EXES) [12-18] with the purpose to detect a specific interaction between the first monomer unit of the chain and the metal responsible of the strong PAN/metal adhesion. According to electrochemical impedance measurements [18], the metal surface is more efficiently blocked by electrodeposited PAN than by a solvent cast PAN film. Formation of C-Ni bond has also been claimed on the basis of UPS and XPS data [14,15]. Interaction between AN and a non noble metal has been discussed on the basis of density functional theory and careful XPS and UPS experiments [19]. Acrylonitrile is chemisorbed on iron, nickel and copper polycrystalline surfaces via the carbon and the nitrogen atoms. Depending on the experimental conditions, the monomer is adsorbed flat on the surface and chemically bound by a $(2p\pi)$-$(3d/4s)$ overlap in which both the C=C double bond and the C≡N triple bond of AN are involved, or it is adsorbed perpendicular to the surface by a covalent interaction between the nitrogen lone pair of AN and the 3d/4s of the metal. The amount of charge transferred to the chemisorbed molecules is increased and increases linearly with the polarization of the metal electrode [20].

1.4 Extension to (Meth)acrylic monomers other than AN

According to Baute et al. [21], the monomer and the solvent compete for being adsorbed at the surface. The preferential adsorption of the monomer is a prerequisite for the electrografting to be successful, which emphasizes that the choice of the solvent is crucial. In this respect, the donicity of the solvent is a very helpful guideline because a solvent of increasing donor number (DN) has a lower ability to adsorb on the cathode. As an example, the electrografting of poly(ethylacrylate) systematically fails in ACN, a solvent with a low donor number (DN=14.1) whatever the monomer concentration, although it occurs successfully in DMF with a higher DN (26.6). So, ethyl acrylate (EA) can displace DMF but not ACN from the metal surface. Consistently, addition of small amounts of ACN keeping constant the EA concentration in a DMF solution, makes the grafting of EA much less efficient. These experimental observations have also been supported by theoretical calculation of the binding energies of various monomers and solvents with a Ni cluster [22]. Clearly, AN has the highest affinity for Ni in agreement with the strong propensity of this monomer to electrografting whatever the solvent used.

Table 1: Relation between the (meth)acrylate monomers and the solvent with the lower DN in which they can be electrografted

DN	Solvent (Monom. conc.)	Acrylic monomers	Methacrylic monomers
0	1,2 DCE (> 1M)	$CH_2{=}CH{-}CN$	
14.1	ACN		
26.6	DMF < 0.5M	$CH_2{=}CH{-}COOEt$, $CH_2{=}CH{-}COOBut$	$CH_2{=}C(CH_3){-}COO{-}(CH_2)_2{-}O{-}Si(Me)_3$
	> 1M		$CH_2{=}C(CH_3){-}COO{-}CH_2{-}CH(O)$ (epoxy)
	> 2M	$CH_2{=}CH{-}COO{-}tBut$	$CH_2{=}C(CH_3){-}COOMe$, $CH_2{=}C(CH_3){-}COO{-}(CH_2)_3{-}Si(OCH_3)_3$
	> 3M		$CH_2{=}C(CH_3){-}COO{-}CH_2{-}CF_3$
33.1 38.8	Py HMPA		$CH_2{=}C(CH_3){-}COO{-}(CH_2)_2{-}N(CH_3)_2$

The solvents that can be used in electropolymerization, are listed in Table 1 according to increasing DN. The monomers are reported facing the solvent of the lower donor number (and at the lower concentration, if meaningful) in which the electrografting of the parent polymer is successful. Chemisorption is observed in any other solvent of a higher DN even at low concentration. Therefore, the appropriate tuning of the competitive adsorption of the monomer and solvent onto the cathode by means of the donor number of the solvent and the monomer concentration, makes it possible to chemisorb successfully a series of poly(meth)acrylates. For instance, polymers containing groups (protected or not) hydrophilic and/or reactive towards various types of topcoats (epoxy, protected carboxylic acid and hydroxyl groups) [21] can be chemisorbed and partly fluorinated polymers (thus anti-adhesive coating) [23], as well.

In a further step, two monomers have been electrografted either simultaneously [24], or in a sequential manner [25]. In the former approach, copolymer brushes are grown from the surface, with a composition that depends on both the comonomer feed composition and the preferential adsorption of the comonomers. In the second approach, a mixed brush is formed on the surface, only when the monomer with the higher reduction potential is polymerized first. Otherwise, the first brushes is degrafted when the second one is built up. As result of this substantial progress, a large variety of functional surfaces can be made available now.

1.5 Electrografting on a Variety of Conducting Substrates

Electrografting has been originally reported in case of common metals such as Fe, Ni, and Cu. It is crucial that these oxidizable metals are pretreated for removing any oxide from the surface. These superficial oxides are actually cathodically reduced in an acetonitrile/TEAP solution, before the transfer of the reduced metal, under inert atmosphere, to the electrochemical cell that contains the monomer to be electrografted. Electrografting onto Zn, Al and Si [26-28], is quite a problem because the superficial oxide layer resists electrochemical reduction. Noble metals, such as Au and Pt, are good candidates for electrografting without preliminary electrochemical treatment. Electrografting has also been reported on various alloys, e.g. stainless steel, brass, nitinol and memory-shape alloys.

Interestingly, electrografting of (meth)acrylates has been reported onto several carbon allotropic forms [29] (at least to (semi)-conducting carbon objects), the chemisorption resulting then from a carbon-carbon bonding. Extension to carbon fibers [30] and carbon black powders [31] has also been found successful. Very recently, electrografting onto multi-walled carbon nanotubes [32] been carried out successfully so making the nanotubes dispersable in a good solvent for the grafted chains.

2. Recent Advances in Electrografting: Toward Stimuli-Responsive Coatings

The judicious choice of the electrografting conditions (solvent, concentration, potential) is very instrumental to impart long term surface modifications to conducting inorganic substrates, including protection, adhesion, and post-reactivity. Nevertheless, the thickness of the organic film is systematically smaller than 200nm. Moreover, although the family of the traditional poly(meth)acrylates that can be chemisorbed on a cathode is large (Table 1), it is highly desirable to increase the range of the chemical properties of the grafted films. In this purpose, new acrylates have been synthesized with the organic function in the ester group, which is ,i.e., a monomer polymerizable after the polyacrylate electrografting [33-34] (pyrrole or thiophene: scheme 2A and norbornene: scheme 2B), an initiator of controlled radical polymerization [35-36] (activated chloride: scheme 2C and alkoxyamine: scheme 2D), an activated ester [37] (prone to nucleophilic substitution: scheme 2E), or a macromonomer (scheme 2F) [38].

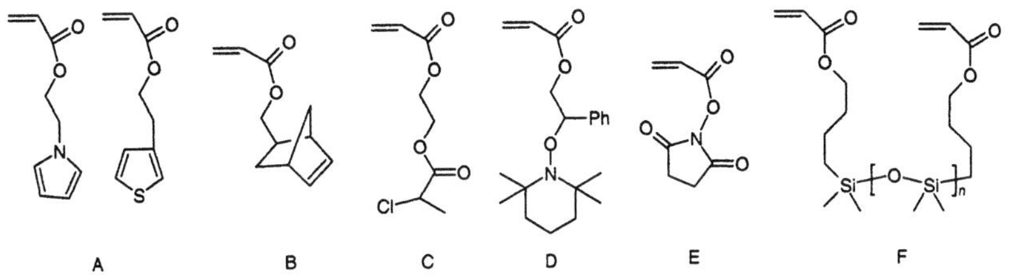

Scheme 2: Structure of new functional acrylates suited to electrografting

2.1 Electrografting of Acrylates Precursor of Conducting Polymers

Electrooxidation of monomers, such as pyrrole and thiophene, is a classical technique to prepare intrinsically conducting polymers. A drawback of this method is the poor adhesion of the film to the substrate. Now, this problem can be solved by the cathodic electrografting of an acrylate that contains a pyrrole or a thiophene unit in the ester group, i. e., N-(2-acryloyloxyethyl) pyrrole (PyA) and 3-(2-acryloyloxyethyl) thiophene (ThiA) (Scheme 2A). The parent polyacrylates are easily chemisorbed under cathodic polarization, followed by a polarization inversion in order to polymerize the pyrrole or thiophene substituents so making the strongly adhering film electrically conducting. Pyrrole or thiophene can be added to the electrochemical bath with the purpose to increase the thickness of the electrically conducting film [33,39].

As an alternative, PyA and ThiA can be polymerized by conventional or controlled radical polymerization. The preformed polyacrylates, polyPyA and polyThiA, are then dissolved in the electrochemical bath or cast onto electrografted APy and AThi, prior to electrooxidation. Depending on the anodic polarization time, electroactive films of a controllable thickness, (up to several microns), can be prepared. This substantial progress is expected to improve the performances of various high tech devices, such as light emitting diodes, anticorrosion coatings, electrochromic windows, and electrochemical sensors, which require the stable and adherent coating of a conducting substrate by a conjugated polymer.

Binary polymer films consisting of an insulating polymer and a conducting polymer have also been prepared by sequential electropolymerization of the parent monomers [29]. The insulating polymer (polyacrylonitrile or polyethylacrylate) was electrografted under cathodic polarization, onto nickel and carbon, respectively. The conducting polymer (polybithiophene or polypyrrole) was deposited in a second step, by electrooxidation of the related monomer in a good solvent for the electrografted chains. In this strategy, only commercially available monomers are advantageously used. Moreover, the adhesion of the conducting polymer is substantially improved, more likely by a chain entrapment effect. Finally, the proper combination of the insulating and the conducting polymers (e.g., PEA and PPy) allows a solvent-responsive film to be prepared. Indeed, the electroactivity of PPy is strongly dependent on the swelling degree of the electrografted P EA c hains. B y c ontrolling t he amount of deposited PPy by the polarization time, only small nuclei that remain below the PEA tethered chains, can be grown on the electrode surface. When PPy electroactivity is recorded in a good solvent for the PEA chains, the high swelling degree of these chains allow the diffusion of ions through the coating and a st rong redox signal typical for PPy is observed (curve a and b Figure 2B). However, in a poor solvent for PEA (for example in water, Figure 2B curve c), the PEA chains precipitate on top of the PPy nuclei and the electroactivity of PPy then completely disappears (Figure 2). These films have thus a potential in solvent sensing devices.

Finally, when the conjugated polymer is synthesized under galvanostatic rather than potentiostatic conditions, the grafted polymer layer acts as a template for the PPy growth [40-43]. In that c ase, p olypyrrole n anowires have been successfully prepared (Figure 3).

2.2 Electrografting of inimers or production of bilayered films

An inimer is a compound t hat contains a monomer and an initiator in the same molecule. In this work, the monomer is an acrylate and the initiator of controlled/living polymerization of (non) (meth)acrylate monomers is part of the ester group. So after the electrografting of the polyacrylate, a second family of chains c an be initiated and grown from the strongly adhering polyacrylate film Therefore the chemical properties of the organic coating can be changed extensively and the film thickness can be increased at will because the second polymerization step is controlled/living while, last but not least, the adhesion is preserved.

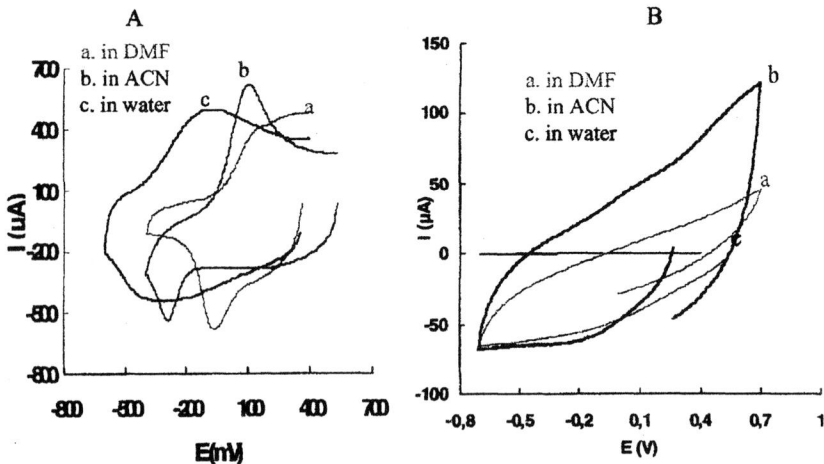

Figure 2: Electroactivity of A) a PPy film B) a PEA/PPy binary film in a) DMF b) ACN c) water.

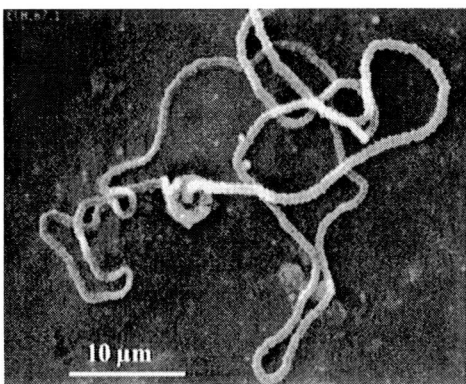

Figure 3: SEM image of a polypyrrole wire grown through a grafted PEA template

Sequential Electrografting and Ring-Opening Metathesis Polymerization (ROMP) of Norbornene and Derivatives

The acrylic acid bicyclo [2.2.1]hept-2-ylmethyl ester (norbornenyl acrylate NBE-A) (Scheme 2B) has been first electrografted, followed by the addition of the Grubbs catalyst $(RuCl_2(=CHPh)-(PCy_3)_2)$ to the pendent double bonds of the chemisorbed chains with formation of ruthenium containing species commonly used to initiate the polymerization of norbornene by ROMP [34]. Because this polymerization is living [44], the thickness of the polynorbornene film can be largely controlled by the monomer concentration and the reaction time.

In addition to an easy implementation, the versatility of this strategy is worth being noted. Indeed, either the double bonds of the polynorbornene chains can be derivatized into reactive groups, such as epoxide, alcohol, amine,...[45], or norbornene can be functionalized with carboxylic acid, anhydride, hydroxyl, cyclodextrine,... groups [46] prior to polymerization, which allows both the film thickness and the surface properties and reactivity to be tuned on purpose.

Sequential Electrografting and Ring-Opening Polymerization (ROP) of Lactides and Lactones.

The electrografting of silylated 2-hydroxyethyl methacrylate has been first carried out with the purpose to make hydrophilic hydroxyl groups available at the film surface after deprotection. These groups have then been reacted with triethyl aluminum and converted into aluminum alkoxide initiators for the ROP of lactides and lactones [47]. As a more direct alternative, the Al alkoxides can be formed by reduction of the ester groups of electrografted polyethylacrylate chains with lithium diisobutylaluminum hydride. The ring-opening polymerization of ε-caprolactone has then been initiated by the alkoxides attached to the primary coating, with the easy control of molecular weight and thus thickness [48].

The brush of poly(ε-caprolactone) chains is an interesting model for investigating crystallization confined in thin films [49]. Coating of metallic implants of various shapes by a polymer known for biocompatibility also deserves attention [50].

Sequential Electrografting and Atom Transfer Radical Polymerization (ATRP)

Electrografting of an acrylate bearing an activated carbon-chloride bond in the ester (Scheme 2C), has been carried out in DMF [35]. Although the stability of the carbon-chloride bond is not ideal under cathodic polarization, the adhering polyacrylate film deposited on the cathode contains unmodified C-Cl bonds (>25%) of the original species. In the presence of a suitable catalyst (e.g., CuCl) the chlorides available at the surface have the capacity to initiate the ATRP of various monomers, such as styrene, n-butyl acrylate and methyl methacrylate. The metallic catalyst must be chosen for compatibility with the conducting substrate. For instance, the $CuCl/CuCl_2$ catalyst commonly used in ATRP [51] is prohibited whenever steel is the cathode because of oxidation by copper (II). Therefore, CuCl has been replaced by the Grubbs catalyst which is inert towards iron. CuCl could however be used with a conductive carbon substrate.

A large variety of substrates (metals, carbon, ...) can be coated now by organic films thicker than 200nm, i.e., the upper limit for electrografting. Monomers such as styrene and vinylpyridine, that cannot be directly electrografted to the cathode, can be now immobilized at the surface with a strong adhesion.

An acrylate bearing an electroactive species (e.g., ferrocene) in the ester could also be polymerized by ATRP from the primary coating. Needless to say that the second ATRP step is not restricted to homopolymerization, but easily extended to random, gradient and block copolymerization if desired.

Sequential Electrografting and Nitroxide-Mediated Polymerization (NMP)

Another type of inimer associates an (electro)polymerizable acrylate and an alkoxyamine, which is the precursor of a pair of initiator/nitroxide mediator for controlled radical polymerization, (Scheme 2D). Polymerization of styrene has been initiated from the chemisorbed polyacrylate chains, and the thickness of the polystyrene layer has been controlled by the molecular weight of the chains (in relation to the amount of an alkoxyamine added to the styrene solution (2-phenyl-2-(2,2,6,6-tetramethyl-piperidin-1-yloxy)-ethylbenzoate PTEMPO) [36]. The polystyrene chains have been recovered by the selective hydrolysis of the ester bond between them and the polyacrylate sublayer. According to SEC analysis, the molecular weight distribution is narrow, and the molecular weights of the grafted chains and the chains grown in solution are in good agreement.

Similarly, a brush of random copolymer of 2-(dimethylamino ethyl)acrylate (DMAEA) and either styrene or n-butyl acrylate [52-53] has been grafted from the polyacrylate sublayer by nitroxide mediated radical polymerization. The strong adhesion of the grafted copolymer chains at the surface of stainless steel has been confirmed by peeling tests. After quaternization of the DMAEA units, the coating exhibits antibacterial properties, as assessed by quartz crystal microbalance experiments. Indeed, the fibrinogen adhesion is two-three times lower after than before quaternization. Moreover, a high antibacterial activity has been measured against Gram-positive bacteria *S. Aureus* and Gram-negative bacteria *E. Coli* for the quaternized coating. As an example, Sty_{62}-$DMAEA_{38}$ copolymer chains are able to completely kill E. Coli after 30min. when quaternized with bromooctane.

Because the chains initiated by NMP retain their capacity of growing for a long time, block copolymers can be grafted from the alkoxyamine containing chemisorbed polyacrylate layer. For instance, an electrografted stainless steel plate has been immersed into n-butylacrylate containing an alkoxyamine, i.e (2,2,5-trimethyl-3-(1'-phenylethoxy)-4-phenyl-3-azahexane) and the parent nitroxide and heated at 125°C for 24h [38]. After careful washing by toluene, the stainless steel substrate has been dipped in styrene added with the PTEMPO alkoxyamine and heated at 125°C for 18h. The block copolymer chains tethered at the surface form nanostructures which are solvent-responsive. Indeed, when immersed in THF, a good solvent for the two blocks, no specific nanostructure is observed by AFM in the phase contrast mode (Figure 4A). Nevertheless, when the film is immersed in n-butanol, a selective solvent for the n-butyl acrylate block, the nanostructuration of the surface is quite obvious (Figure 4B) and results from the selective swelling of one type of nanodomains and precipitation of the other one. Quite interestingly, the swollen nanodomains are potential nanoreactors that can be addressed electrochemically. This opportunity has been illustrated by swelling the P(n-BuA) nanodomains with a saturated solution of copper (II) acetate and tetrabutylammonium tetrafluoroborate (a conducting salt) in 1-butanol. Under potentiostatic regime (-2V/Pt), Cu(II) is reduced and the metal selectively grown in the P(n-BuA) nanodomains. AFM observations confirm that the copper deposit faithfully reproduces the original pattern of the block copolymer film before metal deposition. The pattern formed by copper deposition can actually be tuned by the copolymer composition. This technique is thus a valuable bottom-up approach to design nanostructures (of a metal, a PCE,...) by localized electrodeposition.

2.3 Electrografting of Anchoring Groups

Succinimidyl acrylate is an easily prepared monomer which is well-suited to the surface functionalization of an electrically conducting substrate by succinimidyl groups, which have high propensity to react with nucleophiles, such as amines. The succinimidyl pendant groups of the chemisorbed polyacrylate chains are thus valuable anchoring sites for any type of amine containing (macro)molecules (Scheme 2E) [37-38]. As an example, polyethylene imine chains have been covalently bonded to the polyacrylate film, providing strong anti-bacterial activity to the coating. This grafting technique has also been extended to nucleotides and proteins under mild conditions. Attachment of glucose oxidase enzyme makes the surface sensitive to the addition of glucose. Moreover, biotin-cadaverin has been substituted for the succinimidyl groups at the surface of a quartz crystal microbalance. This

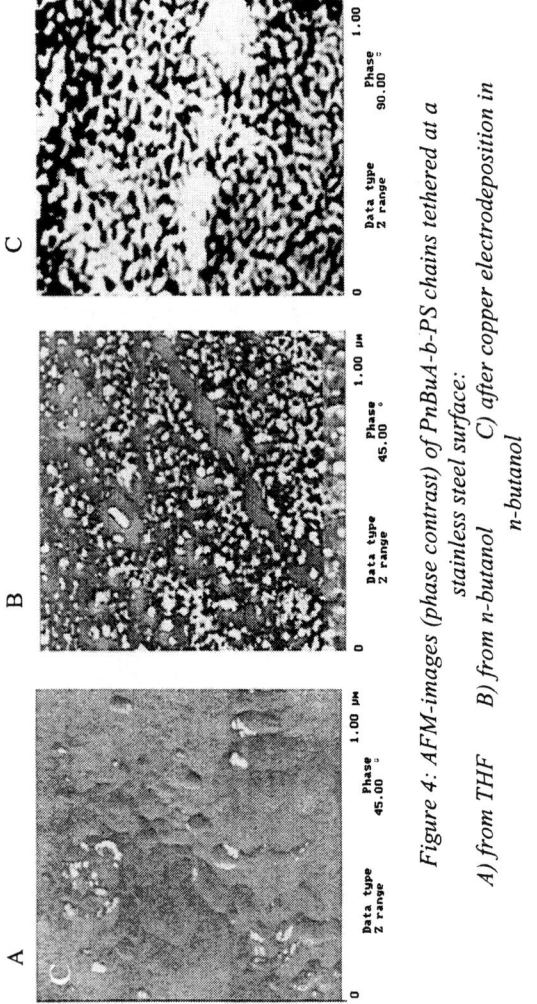

Figure 4: AFM-images (phase contrast) of PnBuA-b-PS chains tethered at a stainless steel surface:

A) from THF   B) from n-butanol   C) after copper electrodeposition in n-butanol

modified surface interacts strongly with avidin and forms a very stable biotin/avidin complex. These examples raise good prospects for the devising of biosensors.

Furthermore, electro-responsive thin films can also be built from poly(succinimidyl acrylate) coatings (Figure 5). Reaction with amino-ferrocene results indeed in electroactive coating. The surface is strongly hydrophobic when iron II is the center of the ferrocene complex. It becomes hydrophilic upon switching the potential anodically as result of the oxidation of iron at the +III state, which makes the film polycationic. The hydrophilicity-hydrophobicity transition is reversible and easily triggered by switching the potential.

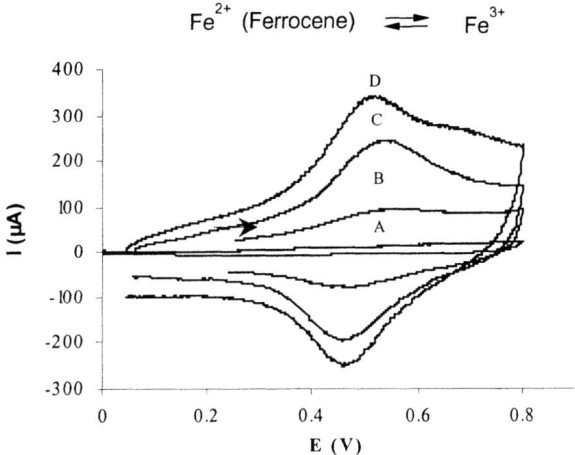

Figure 5: Electroactivity of PNSA films of increasing thickness (B-D) after reaction with aminoferrocene (curve A PNSA before reaction with ferrocene)

2.4 Electrografting of Reactive Polymers

Until recently, the electrografting was reported for small size polymerizable molecules (acrylonitrile and (meth)acrylates, as a rule). Nevertheless, this technique can be extended to preformed macromolecules bearing acrylate substituents, with the advantage that the molecular characteristic features of the chains can be predetermined before chemisorption. In a preliminary study, the electrografting of different polymers of various lengths and contents of acrylate groups has been investigated.

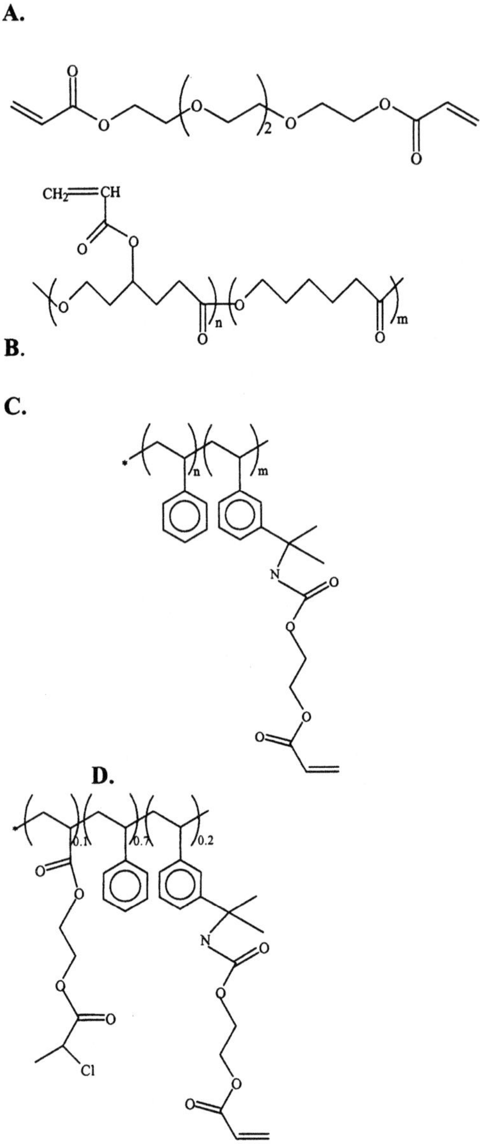

Scheme 3: Examples of acrylate containing polymers suited to electrografting

For instance, oligomers of polyethylene oxide capped by one acrylic unsaturation at one or at both ends have been considered (Scheme 3A) [54]. The electrochemical signature for the grafting, i.e., a passivation peak, has been observed, and the organic coating cannot be removed from the substrate by intensive washing with DMF, a good solvent for the polymer. A priori, monofunctional oligomers (semi-telechelic chains) are expected to form a brush on the surface, whereas telechelic chains would form loops with structural characteristics depending on the chain length and the vicinity of the immobilized chain ends. These films are models for polymer brushes and 2D-crosslinked chains, which are worth being investigated by appropriate analytical experimental techniques. This type of highly hydrophilic coating exhibit protein repulsive properties, as confirmed by comparative data of protein adsorption (BSA and fibrinogen) measured by surface plasmon resonance with a biacore device modified by electrografted PEO monoacrylate or polyethylacrylate, respectively. The adsorption of proteins has been decreased by one order of magnitude in case of PEG-modified surfaces.

Similarly, hydrophobic surfaces have been prepared by direct electrografting of polydimethylsiloxane chains capped by an acrylate at both ends (scheme 2F) [38].

Poly(ε-caprolactone) (PCL) of a well-defined molecular weight and bearing an acrylate substituent per monomer unit (Scheme B with m=0) has been prepared by ROP of the parent dual monomer, γ-acryloxy-ε-caprolactone [55]. Under cathodic polarization, these chains dissolved in DMF (MW=15000, PD=1.1) are chemisorbed onto the electrode [56]. This technique is thus an interesting alternative to the "grafting from" method discussed on §2.2 with the purpose to coat, e.g., a metal by a biocompatible polymer. Random copolymers of ε-caprolactone and the γ-acryloxy-ε-caprolactone (≥50mol%) have also been tested successfully (Scheme 3B). The procedure has been extended to the chemisorption of polystyrene chains containing 15 mol% of acrylates in DMF (Scheme 3C). The presynthesis of "macromonomers" of the acrylate type (with one or more polymerizable units), hydrophobic (polystyrene, polycaprolactone) or hydrophilic (polyethylene oxide), are quite convenient and flexible precursors of organic coatings easily chemisorbed on metals or carbon of various shapes (plates, fibers, etc) [57]. The potential of this novel approach can be substantially increased by using preformed terpolymers. One monomer must promote the desired chemisorption. The second monomer dictates the chemical properties of the envisioned coating (e.g. hydrophobicity / hydrophilicity), and the third comonomer imparts a specific reactivity to the coating (e.g., attachment of molecules of interest, initiation of additional chains, etc.). Scheme 3D shows polystyrene chains prone to be chemisorbed, that contain an initiator of the ATRP of (protected carboxylic acid). Vinylpyridine or protected HEMA can also be used instead of tert-butyl acrylate. In the specific example of chains 3D, a thin hydrophobic layer can be firmly attached to the electrically conducting substrate, whereas a hydrophilic top-layer can be grown from it and used in aqueous media, possibly of biological interest.

The direct grafting of preformed reactive polymers is a very promising strategy to extend further the potential of the electrografting technique. The control of the molecular structure and architecture of premade (co)polymers is a powerful tool for tuning the structure of strongly adhering organic films. In addition to random copolymers and (semi)telechelic polymers, block copolymers, star-shaped (co)polymers, dendrons, etc, of various molecular masses and compositions are candidates for chemisorption. Last but not least, this strategy is applicable not only to polymers prepared by a chain-growth mechanism but also to polycondensates (e.g., after derivatization of side or end-groups into acrylates).

2.5 Applications

Several specific achievements have been reported as a direct application of the AN electrografting, such as preparation of thin α-emitter sources by the simultaneous electroreduction of AN and uranium or neptunium salts [58], preparation of electrical connectors by appropriate thermal post treatment of electrografted PAN thin films [59], metals waste treatment [60] and protection of art pieces [61-62].

A substantial advantage of electrografting is the low current density which makes the process weakly sensitive to ohmic drops. Consequently, any object, whatever the shape, can be homogeneously electrocoated by the chemisorbed polymer, e.g., cardiovascular stents [50], metal plates or wires, carbon fibers or mats [30], etc. Conducting particles or carbon powders have also been coated as result of intermittent contact with a zinc working electrode [31]. Moreover, when a substrate combines both a non-oxidized metal (on which the electrografting is successful) and an oxide (e.g., silicon oxide), the electrografting is selective on the reduced metal [26,63]. This is quite a new and interesting approach of a mask-free patterning of conducting surfaces.

Electrografting has also been extended to AFM-tips in order to make them reactive/functional and to get a better insight on the structural characteristics of the chemisorbed chains and the underlying electrografting phenomenon [64]. Moreover, chains preattached to the tip could be deposited on predefined spots of a surface, under appropriate conditions.

A very large variety of polymer films can be now deposited with strong adhesion onto various substrates by electrochemistry, with the possibility to have a second layer of chains grown from the first one under controlled/living conditions [65-66]. In this case, the electrodeposited film is thin and plays the role of a "primer", whereas the top layer is very versatile in terms of molecular structure, composition, and thickness. The unique flexibility of this new strategy of coating or (multilayer) film deposition raises very optimistic forecasts for protection of metals against corrosion, biocompatibilization of implants, electro-optical applications. Moreover, sizing of fillers by this technique is expected to improve dispersion within polymer matrices and properties the final composites. The grafting of polymers with a well-defined structure onto conducting surfaces and the possible design of these structures paves the way to applications in lithography, nanopatterning, and development of electrochemical sensors.

Acknowledgments

The authors are much indebted to the "Belgium Science Policy" for financial support in the frame of the "Interuniversity Attraction Poles Programme (PAI V/03): Supramolecular Chemistry and Supramolecular Catalysis". C.J. is Research Associate by the "Fonds National de la Recherche Scientifique" (Belgium). The authors are also grateful to S. Gabriel, S. Voccia and Dr. C. Detrembleur for their involvement in part of this work.

References

[1] G. Lécayon, *Chem. Phys. Lett.* **1982**, *91*, 506; C. Boiziau, G. Lécayon *La Recherche*, **1988**, *19*, 888; C. Boiziau, G. Lécayon, *Surface and Interface Analysis*, **1988**, 475
[2] N. Baute, C. Calberg, P. Dubois, C. Jérôme, R. Jérôme, L. Martinot, M. Mertens, P. Teyssie, *Macromolecular Symposia* **1998**, *134* (Electron Transfer Processes and Reactive Intermediates in Modern Chemistry), 157
[3] N. Baute, L. Martinot, R. Jérôme, *Journal of Electroanalytical Chemistry* **1999**, *472(1)*, 83
[4] C. Calberg, M. Mertens, R. Jérôme, X. Arys, A. M. Jonas, R. Legras, *Thin Solid Films*, **1997**, *310*, 148
[5] N. Baute, C. Jérôme, L. Martinot, M. Mertens, V. Geskin, R. Lazzaroni, J.L. Brédas, R. Jérôme, *European Journal of Inorganic Chemistry* **2001**, *5*, 1097
[6] C. Boiziau, S. Leroy, C. Reynaud, G. Lécayon, C. Le Gressus, P. Viel, *J. Adhesion*, **1987**, *23*, 21; D.E. Labaye, C. Jérôme, R. Jérôme, *Polymeric Materials Science and Engineering* **1999**, *80*, 10
[7] C. Bureau, G. Deniau, P. Viel, G. Lécayon, *Macromolecules*, **1997**, 30, 333
[8] V. Geskin, R. Lazzaroni, M. Mertens, R. Jérôme, J.L. Brédas, *J. Chem. Phys.*, **1996**, 105, 3278
[9] N. Baute: Ph. D. Thesis, University of Liège, **1999**, Chapter 5
[10] J. Charlier, C. Bureau, G. Lécayon, *J. Electroanal. Chem.* **1999**, 465, 200
[11] J. Tanguy, *J. Electroanal. Chem.* **2000**, *487*, 120
[12] P. Viel, S. De Cayeux, G. Lécayon, *Surf. Interf. Anal.*, **1993**, 20, 468
[13] D. Mathieu, M. Defranceschy, P. Viel, G. Lécayon, J. Delhalle, *Polym. Solid Interf., Proc. Int. Conf.*, 1st, Institute of Physics, Bristol, **1992**, 379
[14] Y. Bouizem, F. Chao, M. Costa, A. Tadjedinne, G. Lécayon, *J. Electroanal. Chem.*, **1984**, 172, 101
[15] P. Jonnard, F. Vergand, P.F. Staub, C. Bonnelle, G. Deniau, C. Bureau, G. Lécayon, *Surf. Interface Anal.*, **1996**, *24*, 339
[16] X. Crispin, V. Geskin, A. Crispin, J. Cornil, R. Lazzaroni, W. R. Salaneck, J. L. Brédas, *J. Am. Chem. Soc.*, **2002**, *124(27)*, 8131
[17] X. Crispin, C. Bureau, V. Geskin, R. Lazzaroni, W. R. Salaneck, J. L. Brédas, *J. Chem. Phys.*, **1999**, *111(7)*, 3237

[18] J. Tanguy, P. Viel, G. Deniau, G. Lécayon, *Electrochim. Acta*, **1993**, *38*, 1501
[19] X. Crispin, R. Lazzaroni, A. Crispin, V. M. Geskin, J. L. Bredas, W. R. Salaneck, *Journal of Electron Spectroscopy and Related Phenomena* **2001**, *121(1-3)*, 57
[20] X. Crispin, V. Geskin, C. Bureau, R. Lazzaroni, W. Schmickler, J. L. Brédas, *J. Chem. Phys.*, **2001**, *115(22)*, 10493
[21] N. Baute, P. Teyssie, L. Martinot, M. Mertens, P. Dubois, R. Jérôme, *European Journal of Inorganic Chemistry* **1998**, *11*, 1711
[22] X. Crispin, R. Lazzaroni, V. Geskin, N. Baute, P. Dubois, R. Jérôme, J. L. Brédas, *J. Am. Chem. Soc.* **1999**, *121(1)*, 176
[23] H. Serwas, S. Voccia, R. Jérôme, C. Jérôme, to be submitted
[24] N. Baute: Ph. D. Thesis, University of Liège, **1999**, Chapter 3
[25] N. Baute, V. M.Geskin, R. Lazzaroni, J.L. Brédas, X. Arys, A.M. Jonas, R. Legras, C. Poleunis, P. Bertrand, C. Jérôme, R. Jérôme, *e-polymers*, **2004**, n° 063
[26] G. Lécayon, A. Le Moël, C. Le Gressus; *Vide, les couches minces*, **1980**, *201*, 638
[27] C. Fredrikson, R. Lazzaroni, J.L. Brédas, M. Mertens, R. Jérôme, *Chem. Phys. Lett.*, **1996**, *258*, 356.
[28] J. Charlier, S. Ameur, J.P. Bourgoin, C. Bureau, S. Palacin, *Advanced Functional Materials* **2004**, *14(2)*, 125-132.
[29] C. Jérôme, V. Geskin, R. Lazzaroni, J. L. Brédas, A. Thibaut, C. Calberg, I. Bodart, M. Mertens, L. Martinot, D. Rodrigue, J. Riga, R. Jérôme, *Chem. Mater.* **2001**, 13(5), 1656-1664
[30] M. Mertens, L. Martinot, R. Jérôme, *PCT Int. Appl. WO 9902614*, **1999**, 27 pp.
[31] R. Jérôme, L. Martinot, M. Mertens, *PCT Int. Appl. WO 9920697*, **1999**, 30 pp.
[32] P. Petrov, X. Lou, C. Pagnoulle, C. Jérôme, C. Calberg, R. Jérôme, *Macromolecular Rapid Communications* **2004**, *25(10)*, 987
[33] D. E. Labaye, C. Jérôme, V. M. Geskin, P. Louette, R. Lazzaroni, L. Martinot, R. Jérôme, *Langmuir* **2002**, *18(13)*, 5222
[34] C. Detrembleur, C. Jérôme, M. Claes, P. Louette, R. Jérôme, *Angew. Chem., Int. Ed. Engl.* **2001**, *40(7)*, 1268
[35] M. Claes, S. Voccia, C. Detrembleur, C. Jérôme, B. Gilbert, Ph. Leclère, V. M. Geskin, R. Gouttebaron, M. Hecq, R. Lazzaroni, R. Jérôme, *Macromolecules* **2003**, *36(16)*, 5926
[36] S. Voccia, C. Jérôme, C. Detrembleur, Ph. Leclère, R. Gouttebaron, M. Hecq, B. Gilbert, R. Lazzaroni, R. Jérôme, *Chem. Mater.* **2003**, *15(4)*, 923
[37] C. Jérôme, S. Gabriel, S. Voccia, C. Detrembleur, M. Ignatova, R. Gouttebaron, R. Jérôme, *J. Chem. Soc. Chem. Commun.* **2003**, *19*, 2500
[38] R. Jérôme, S. Gabriel, S. Voccia, C. Jérôme, *Polymeric Materials Science and Engineering* **2004**, *90*, 218
[39] D. E. Labaye, C. Jérôme, R. Jérôme, *Polymeric Materials Science and Engineering* **1999**, *80*, 10
[40] C. Jérôme, R. Jérôme, *Angew. Chem., Int. Ed. Engl.* **1998**, 37(18), 2488
[41] C. Jérôme, D. Labaye, I. Bodart, R. Jérôme, *Synthetic Metals* **1999**, *101(1-3)*, 3

[42] C. Jérôme, S. Demoustier-Champagne, R. Legras, R. Jérôme, *Chemistry--A European Journal* **2000**, *6(17)*, 3089
[43] C. Jérôme, D. E. Labaye, R. Jérôme, *Synthetic Metals,* **2004**, *142(1-3)*, 207
[44] P. Schwab, M.B. France, J.W. Ziller, R.H. Grubbs, *Angew. Chem. Int. Ed. Engl.*, **1995**, *107*, 2179
[45] F.M. Sinner, M.R. Buchmeiser, *Angew. Chem. Int. Ed.*, **2000**, *39*, 1433
[46] S. Voccia, M. Claes, R. Jérôme, C. Jérôme, accepted in *Macromolecular Rapid Communications.*
[47] A. Löfgren, A. C. Albertsson, P. Dubois, R. Jérôme, *J. Macromol. Sci. - Rev. Macromol. Chem. Phys.*, **1995**, *C35(3)*, 379-418
[48] S. Voccia, L. Beck, B. Gilbert, R. Jérôme, C. Jérôme, *Langmuir*, **2004**, *20*, 10670
[49] C. Celles, J.P. Chapel, S.Voccia, R. Jérôme, C. Jérôme, *Abstract book of IUPAC* **2004**, *Paris.*
[50] C. Jérôme, A. Aqil, S. Voccia, D.E. Labaye, V. Maquet, S. Gautier, O. F. Bertrand, R. Jérôme, *J. Biomed. Mat. Res.*, accepted
[51] K. Matyjaszewski, P.J. Miller, N. Shukla, B. Immaraparn, A. Gelman, B. Luokala, T. Siclovan, G. Kickellick, T. Vallant, H. Hoffmann, T. Pakula, *Macromolecules*, **1999**, *32*, 8716
[52] C. Jérôme, M. Ignatova, S. Voccia, S. Lenoir, R. Jérôme, *Polymeric Materials Science and Engineering,* **2004**, *90*, 109
[53] M. Ignatova, S. Voccia, B. Gilbert, N. Markova, P. S. Mercuri, M. Galleni, V. Sciannamea, S. Lenoir, D. Cossement, R. Gouttebaron, R. Jérôme, C. Jérôme, *Langmuir*, **2004**, *20*, 10718
[54] S. Gabriel, P. Dubruel, E. Schacht, A. Jonas, B. Gilbert, R. Jérôme, C. Jérôme, *Angew. Chem., Int. Ed. Engl.*, submitted
[55] D. Mecerreyes, J. Humes, R.D. Miller, J.L. Hedrick, C. Detrembleur, P. Lecomte, R. Jérôme, J. San Roman, *Macromol. Rapid Commun.,* **2000**, *21*, 779
[56] X. Lou, C. Jérôme, C. Detrembleur, R. Jérôme, *Langmuir* **2002**, *18(7)*, 2785
[57] C. Jérôme, M. Claes, C. Detrembleur, R. Jérôme, to be submitted to *Polymer*
[58] L. Martinot, M. Mertens, L. Lopes, P. Faack, M. Krausch, J. Guillaume, G. Ghitti, J. Marien, J. Riga, R. Jérôme, J. Schrijnemackers; *Radiochimica Acta*, **1996**, *75*, 111-119
[59] F. Houzé, L. Boyer, S. Noël, P. Viel, G. Lécayon, J-M. Bourin, *Synth. Met.*, **1994**, *62*, 207.
[60] C. Bureau, F. Lederf, P. Viel, F. Descours, *PCT Int. Appl. WO 2002018050* **2002**, 105 pp
[61] M. Mertens, C. Calberg, R. Jérôme, L. Martinot, J. Schrijnemackers, J. Guillaume, *Revue des Historiens de l'Art, des Archéologues, des Musicologues et des Orientalistes de l'Université de Liège*, **1997**, 16
[62] J. Guillaume, L. Martinot, S. Gabriel, C. Jérôme, R. Jérôme, G. Weber, M. Merten, E. Duykaerts, *« Art et Chimie, les Polymères »* Actes du Congrès, CNRS Ed., **2003**, Paris
[63] C. Bureau, S. Palacin, J.P. Bourgoin, S. Ameur, J. Charlier, *PCT Int. Appl. WO 2002070148* **2002**, 65 pp
[64] C. Jérôme, N. Willet, R. Jérôme, A.S. Duwez, *ChemPhysChem* **2004**, *5(1)*, 147-149

[65] S. Ameur, C. Bureau, J. Charlier, S. Palacin, *Journal of Physical Chemistry B* **2004**, *108(34)*, 13042

[66] O. Bertrand, R. Jérôme, S. Gautier, V. Maquet, C. Detrembleur, C. Jérôme, S. Voccia, M. Claes, X. Lou, D.E. Labaye, *PCT Int. Appl. WO 2002098926*, **2002**, 43 pp

Physico-Chemical Aspects of Stimuli-Responsive Polymers

Chapter 7

Temperature-Responsive Films of Amphiphilic Poly(*N*-isopropylacrylamides) of Various Architectures at the Air–Water Interface

Roger C. W. Liu and Françoise M. Winnik*

Départment de Chimie et Faculté de Pharmacie, Université de Montréal, C.P. 6128, Succursale Centre-ville Montréal, Québec H3C 3J7, Canada

The surface behavior of hydrophobically modified poly(*N*-isopropylacrylamides) (HM-PNIPAM) was studied by the Langmuir film balance technique in order to explore the role of polymer architecture on the formation and stability of polymeric monolayers. The polymers had similar molecular weight ($M_n \sim 15$ kDa) and narrow polydispersity ($M_w/M_n < 1.1$) and belonged to two general families: (i) graft-PNIPAM derivatives carrying 10 mol% of either *n*-dodecyl chains or fluorinated *n*-octyl chains randomly distributed along the polymer backbone and (ii) end-functionalized PNIPAM derivatives bearing one or two terminal *n*-octadecyl chains. Surface pressure (π)-surface area (A) isotherms were recorded during compression and expansion of polymer monolayers on an aqueous subphase kept at 12 or 20 °C. We observed that hydrophobic modification stabilizes the PNIPAM monolayers formed at the air/water interface, and that telechelic PNIPAM forms the most robust monolayers. Polymeric monolayers formed at 12 °C are less stable than those constructed at 20 °C, reflecting the increased solubility of PNIPAM in cold water.

© 2005 American Chemical Society

The properties of functionalized amphiphilic copolymers have been examined from various viewpoints for their ability to form organized assemblies with different morphologies in solution, in the bulk state, and in thin polymer layers at surfaces and interfaces. The thrust of these studies is wide, encompassing practical applications in drug delivery, the coatings industry, food and cosmetics formulations, as well as theoretical considerations related to solvent/polymer interactions and structure-driven polymer conformations. Among neutral amphiphilic polymers, copolymers of N-isopropylacrylamide (NIPAM) stand out, as they have been the objects of numerous studies over the last three decades. Poly(N-isopropylacrylamide) (PNIPAM) is soluble in cold water, but its aqueous solution undergoes a temperature-induced phase transition as the temperature reaches a critical value ($\sim$ 32 °C) (*1*), at which the polymer conformation changes from a coil to a globule with subsequent self-aggregation. The transformation is reversible and highly reproducible. The phase transition temperature or lower critical solution temperature (LCST) can be modulated by modifying PNIPAM with either polar or non-polar moieties (*2-5*), which perturb the hydrophobic-hydrophilic balance of the polymer chain. Introducing polar moieties results in an increase of the transition temperature, and in some cases, the response to temperature vanishes entirely. In contrast, introduction of non-polar moieties tends to lower the transition temperature and may result in the insolubility of the polymer in water. Important parameters controlling the temperature-sensitivity of NIPAM copolymer solutions include the polymer architecture as well as the type of hydrophobes and the degree of modification. These parameters also affect the association of the polymer in solutions kept below their phase transition temperature. Thus, hydrophobically modified (HM) PNIPAM form micelle-like clusters with multiple hydrophobic microdomains (*6*), end-modified PNIPAM form star-like micelles (*7*), whereas telechelic PNIPAM derivatives assemble in flower-like micelles (*8*).

The surface properties of PNIPAM and its derivatives have been the subject of a number of studies, via static methods (*9-11*), such as tensiometry, neutron reflectivity, and ellipsometry, to detect the presence of a polymer layer at the air/water interface and to measure the thickness of the layer. The dynamics of PNIPAM at the air/water interface were tested by surface laser light scattering, oscillating pendant drop techniques, Langmuir trough experiments, and time-resolved null-ellipsometry (*12-14*). The surface tension of PNIPAM solutions, determined by the Whilhelmy method (*15*), is $\sim$ 42 mJ m^{-2} at room temperature and $\sim$ 48 mJ m^{-2} at 31 °C. Oscillating pendant drop experiments (*16*), probing the dynamics of surface rearrangement, revealed small surface tension oscillations, attributed to fast loop-to-train transitions during surface expansion. Using a Langmuir trough and ellipsometry (*17-19*), Kawaguchi and coworkers measured pressure-induced changes in polymer surface concentration and in the thickness of the adsorbed polymer layer. They reported coverage of about 2 mg m^{-2} for solutions kept at 16 °C. From these studies, it was inferred that the adsorption of PNIPAM at the air/water interface consists of two steps: (i) a fast adsorption leading to a layer of chains laying mainly flat, creating a two-dimensional surface film characterized by a high dynamic surface elasticity, and (ii) formation of long loops and tails, leading to faster relaxation processes and, consequently, to a decrease of the dynamic surface elasticity.

The rich interfacial properties of PNIPAM in water have fascinated experimentalists and theoreticians alike, but from the practical standpoint, the fragility of the PNIPAM monolayer is an insurmountable drawback. Poly(N-n-alkylacrylamides) with n-alkyl chains ranging in length from six (n-hexyl) to eighteen (n-octadecyl) carbons form much more robust monolayers, stabilized via hydrophobic interactions between the alkyl groups and via hydrogen bonds between the amide groups and water molecules of the subphase (20-22). Among the various poly(N-n-alkylacrylamides), poly(N-n-dodecylacrylamide) forms the most stable monolayers (23). Monolayers of poly(N-alkylacrylamides) with non-linear alkyl groups, including the *tert*-pentyl and adamantyl moieties, also form monolayers at the air/water interface, albeit of reduced stability, compared to linear chains (24-26). Fluorinated poly(N-n-alkylacrylamides), with substituents such as 2-(n-perfluorooctyl)ethyl, form monolayers as well (27). Their properties depend strongly on the length of the fluorocarbon chains.

The disparity between the properties of PNIPAM monolayers and their poly(N-n-alkylacrylamide) counterparts suggests that HM-PNIPAM may present an intermediate interfacial behavior and, more importantly, that it may be possible to create thin films of controlled properties via careful molecular design. There have been very few studies of the interfacial properties of PNIPAM end-functionalized or grafted with n-alkyl chains, particularly, with high levels of hydrophobic modification. Of note is a report of Shi and coworkers on the properties of a copolymer of NIPAM and n-octadecyl acrylate (molar ratio 64:1), which indicates that this copolymer forms monolayers at the air-water interface (28). Interestingly, these researchers noted an unexpected temperature response of the monolayers, for temperatures ranging between 10 and 25 °C, temperatures lower than the LCST of PNIPAM. Intrigued by this report we set about to study systematically the interfacial properties at the air/water interface of HM-PNIPAM of controlled molecular characteristics (29). The HM-PNIPAM samples were prepared by radical addition-fragmentation chain transfer (RAFT) polymerization, a living free radical polymerization technique that allows the preparation of polymers with narrow molecular weight distribution and telechelic functionality (30). Monolayers formed by the copolymers at the air/water interface were investigated as a function of polymer architecture, in order to develop an understanding of the polymer structure-property relationship. Surface pressure (π) surface area (A) isotherms were obtained for two subphase temperatures, 12 and 20 °C, to investigate the temperature responsiveness of HM-PNIPAM monolayers below the LCST of PNIPAM itself.

Architecture of the HM-PNIPAM Samples Under Scrutiny

Four HM-PNIPAM samples of similar molecular weights ($M_n \sim 15\,000$ Da) but different architecture, were selected: (i) two graft-PNIPAM derivatives bearing a modest level (10 mol%) of n-dodecyl or fluorinated n-octyl chains (PNIPAM-H and PNIPAM-F, respectively, Figure 1), (ii) an end-functionalized PNIPAM derivative carrying an n-octadecyl group at one chain end and an isopropyl group at the other (C_{18}-PNIPAM-C_3, Figure 1), and (iii) a telechelic PNIPAM derivative with an n-octadecyl group at each chain end (C_{18}-PNIPAM-C_{18}, Figure 1). The graft-PNIPAM derivatives were synthesized by RAFT copolymerization (dioxane, 65 °C) of NIPAM and either N-n-dodecylacrylamide or N-n-1H,1H-perfluorooctylacrylamide using the trithiocarbonate, 2-n-dodecylsulfanylthiocarbonylsulfanyl-2-methylpropionic acid, as RAFT agent and N,N-azobis(isobutyrontrile) (AIBN) as initiator. The PNIPAM sample terminated with an isobutyl group and an isopropyl group (C_4-PNIPAM-C_3), and the telechelic PNIPAM derivative (C_{18}-PNIPAM-C_{18}) were obtained under similar conditions, using the appropriate RAFT agents (8).

Isotherms of Graft-PNIPAM Derivatives

Surface pressure-area isotherms obtained for PNIPAM-F, PNIPAM-H and C_4-PNIPAM-C_3 are presented in Figure 2. A constant volume (10 µl) of solutions of PNIPAM-F (1.15 mg/ml), PNIPAM-H (0.98 mg/ml), or C_4-PNIPAM-C_3 (0.89 mg/ml) in chloroform was spread onto the water surface at a constant subphase temperature (20 ± 0.1 °C). Note that the polymer concentrations were such that in all cases the same number of monomer units was spread onto the aqueous subphase surface. Monolayers were then compressed with a constant compression rate of 20 mm/min (31).

All isotherms show a small increase in surface pressure (< 3 mN/m) upon compression to ~ 0.29 nm^2, followed by a steady increase as the monolayers were further compressed. As the surface reaches ~ 0.17 nm^2, the surface pressure continues to increase in a more gentle way in the case of the graft PNIPAM derivatives, but it levels off in the case of C_4-PNIPAM-C_3 reaching a constant value of ~ 28 mN/m at small surface areas. The surface pressure plateau is diagnostic of the collapse of C_4-PNIPAM-C_3 monolayer, whereas the profiles recorded for PNIPAM-F and PNIPAM-H are indicative of the formation of stable Langmuir films on the water surface (*32*). The enhancement in monolayer stability is slightly more pronounced for PNIPAM-F than PNIPAM-H.

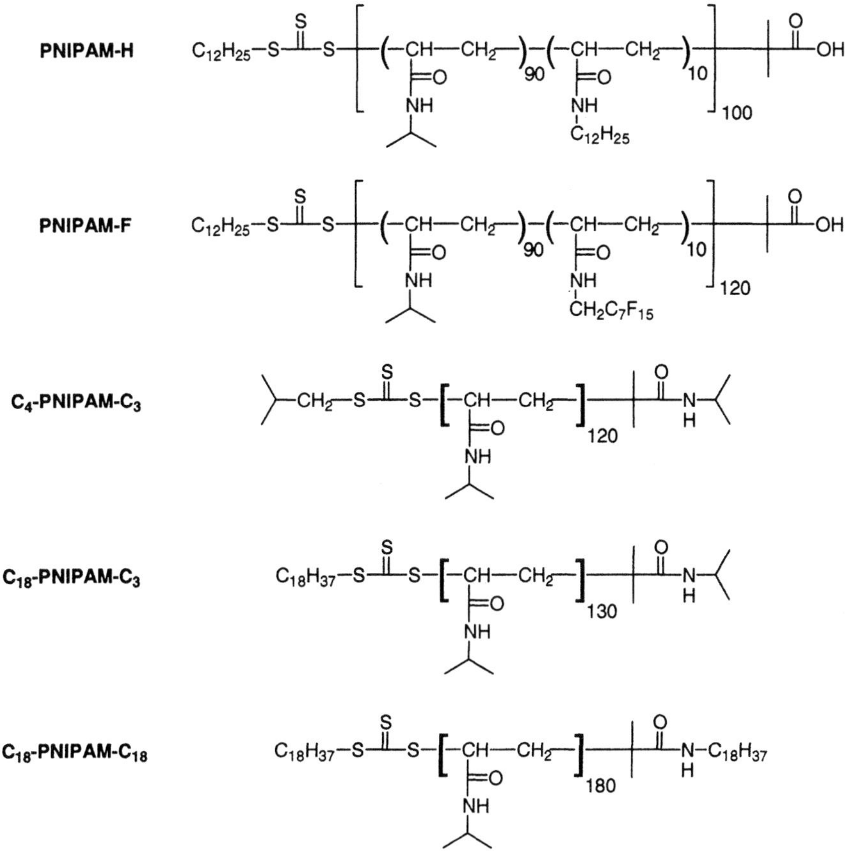

Figure 1. Chemical structure of the PNIPAM Derivatives.

Immediately after compression to the minimum attainable surface area, the monolayers were subjected to a constant-rate expansion (20 mm/min). The expansion isotherms overlapped with the compression isotherms in the cases of PNIPAM-F and PNIPAM-H, confirming the stability and the reversibility upon compression and expansion of the monolayers, with no collapse of the monolayers and/or loss of polymer chains into the subphase (Figure 3). However, hysteresis was observed in the compression-expansion cycles of C_4-

PNIPAM-C$_3$ monolayers. Next, the compression-expansion cycles were acquired with a subphase temperature of 12 °C (Figure 4).

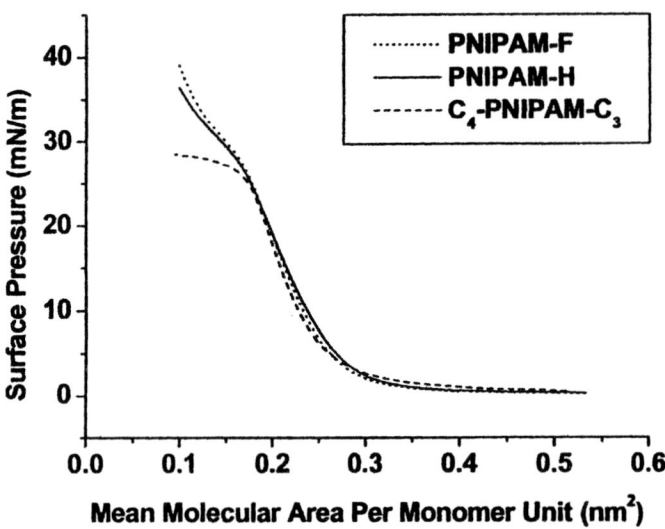

Figure 2. Compression isotherms of the monolayers of PNIPAM-F, PNIPAM-H, and C$_4$-PNIPAM-C$_3$ with a subphase temperature of 20 °C.

A number of revealing trends emerged from these sets of measurements. Turning our attention first to the isotherms registered for C$_4$-PNIPAM-C$_3$, which carries no long *n*-alkyl chain, we note that the hysteresis is more pronounced and the plateau surface pressure is lower at 12 °C (~ 26 mN/m), compared to 20 °C. These trends are consistent with results of previous studies by tensiometry of PNIPAM films spread at the air/water interface that indicated a pronounced tendency of PNIPAM to desorb from the air/water interface and dissolve into the aqueous subphase (*33*). When spread on the water surface, the NIPAM units adsorb at the air/water interface. Compressing the monolayer increases the surface polymer concentration at the interface and, consequently, enhances the repulsive interactions among polymer chains, resulting in a decrease in surface pressure and in the appearance of hysteresis. It is known that the solubility of PNIPAM in water increases with decreasing temperature. The enhanced hysteresis observed at 12 °C is a reflection of the enhanced solubility of C$_4$-PNIPAM-C$_3$, which favors desorption of polymer chains from the interface into the subphase.

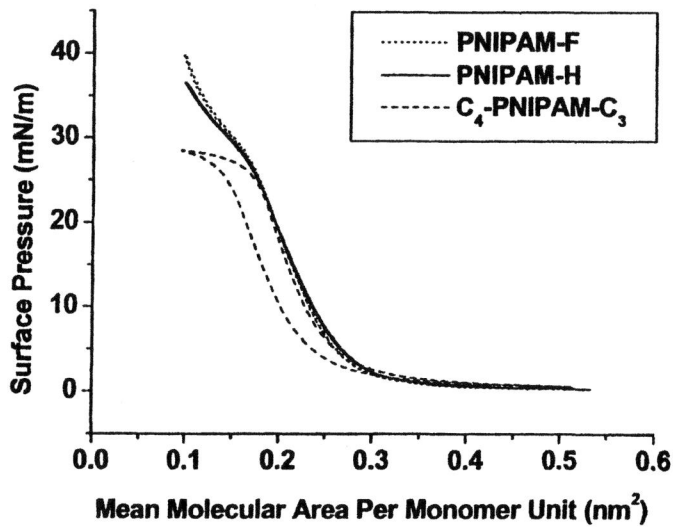

Figure 3. Compression-expansion cycles of monolayers of PNIPAM-F, PNIPAM-H, and C_4-PNIPAM-C_3 (subphase temperature: 20 °C).

Qualitatively, the graft-PNIPAM samples exhibited a similar behavior, albeit much less pronounced, especially in the case of PNIPAM-F. Unlike C_4-PNIPAM-C_3, neither PNIPAM-F nor PNIPAM-H is soluble in water, but in Langmuir films under compression they can adopt a conformation such that the NIPAM segments are in contact with the aqueous subphase. Reducing the surface area and/or the subphase temperature promotes the solubilization of the segments of NIPAM main chain, thus lowering the total number of surface active monomer units at the interface. The solubilization of NIPAM segments, however, is opposed by the preferential orientation of the hydrophobic chains toward the air phase and by the nteractions among the hydrophobic substituents. The hydrophobic chains keep the polymer close to the water surface, allowing a close packing of the polymer chains at the air/water interface that gives a continuous increase in surface pressure under high compression. Fluorinated *n*-octyl chains are more effective, compared to *n*-dodecyl chains, in stabilizing the monolayer against a decrease in subphase temperature, although the fluorinated alkyl chains ($CH_2C_7F_{15}$) are shorter than the alkyl chains ($C_{12}H_{25}$). A CF_2 unit has been shown to be equivalent to 1.7 CH_2 units (*34,35*) in the cases of fluorinated surfactants and fluorocarbon-modified polyelectrolytes bearing small amounts of fluorinated side chains (3 or 7 mol%). The results reported here provide further evidence in support of the $CF_2 = 1.7$ CH_2 relationship, which seems to hold also in monolayers of neutral water insoluble polymers with a modest degree of hydrophobic modification (10 mol%).

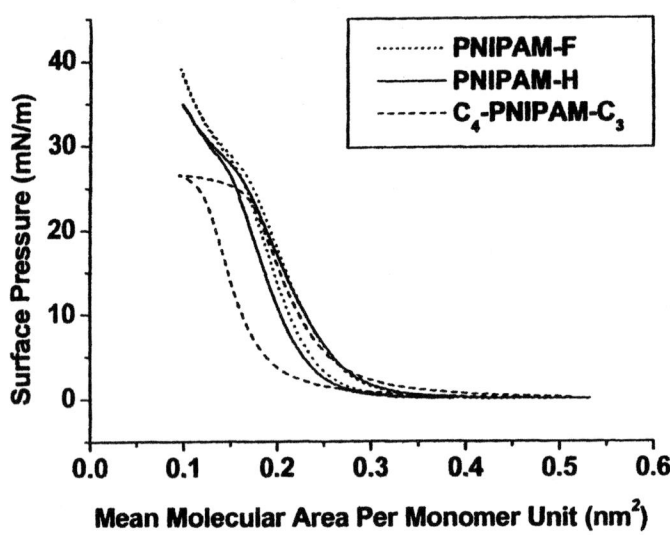

Figure 4. Compression-expansion cycles of the monolayers of PNIPAM-F, PNIPAM-H, and C_4-PNIPAM-C_3 at a subphase temperature of 12 °C.

Taken together, the results provide strong evidence that by modifying PNIPAM with a modest level of hydrophobes (up to 10 mol% in this case), it is possible to enhance significantly the stability of the monolayers without sacrificing the temperature-responsive characteristics of PNIPAM itself.

Isotherms of End-Functionalized PNIPAM Derivatives

Monolayers of the singly end-capped PNIPAM derivative (C_{18}-PNIPAM-C_3) and the telechelic PNIPAM derivative (C_{18}-PNIPAM-C_{18}) were spread on the air/water interface under the same conditions as those used in the case of graft-PNIPAM derivatives, using chloroform solutions (0.9 mg/ml) of each polymer. Isotherms recorded with subphase temperatures of 12 and 20 °C for C_{18}-PNIPAM-C_3 and C_{18}-PNIPAM-C_{18} are presented in Figures 5 and 6. We note that monolayers of singly C_{18}-PNIPAM-C_3 (Figure 5) exhibit a behavior similar to that of C_4-PNIPAM-C_3: as the temperature passes from 20 to 12 °C,

C_{18}-PNIPAM-C_3 becomes less surface active and the hysteresis exhibited by the compression-expansion isotherms increases, albeit to a lesser extent than in the case of the unmodified polymer (C_4-PNIPAM-C_3). In contrast, isotherms recorded during compression-expansion cycles of the telechelic sample (C_{18}-PNIPAM-C_3, Figure 6) present no hysteresis, when the water subphase is kept at 20 °C. A weak hysteresis exists for isotherms recorded at 12 °C, indicating a temperature/compression-induced conformational change of the polymer on the water surface. Thus, introducing a second terminal *n*-octadecyl chain drastically stabilizes monolayers formed at the air/water interface.

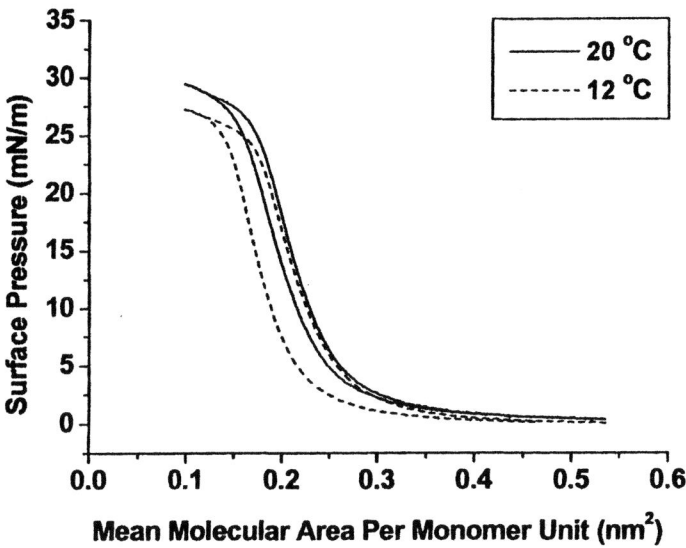

Figure 5. Compression-expansion cycles of the monolayers of C_{18}-PNIPAM-C_3 at two subphase temperatures.

The remarkable increase in monolayer stability, via a slight alteration of the polymer composition and architecture, mirrors changes in the solution properties of the three polymers. The unmodified polymer (C_4-PNIPAM-C_3) is soluble in water below 31 °C, with no trace of aggregation. The end-tagged polymer (C_{18}-PNIPAM-C_3) is also soluble in water below ~ 31 °C, but it forms star-like micelles even under conditions of high dilution. The telechelic sample (C_{18}-PNIPAM-C_{18}), however, appears to be insoluble in water, even below 31 °C, at least as judged by light scattering measurements.

It is interesting to note that, following compression of either C_{18}-PNIPAM-C_{18} or C_{18}-PNIPAM-C_3, at the initial stage of the expansion process, expansion and compression isotherms overlap (Figures 5 and 6); they eventually diverge upon reaching a given surface area, which is lower, in the case of the telechelic polymer, another indication of the enhanced stability of the monolayer formed by C_{18}-PNIPAM-C_{18}, compared to C_{18}-PNIPAM-C_3. For C_4-PNIPAM-C_3, the expansion isotherm negatively deviates from the compression isotherm as soon as the expansion is initiated (Figures 3 and 4). These results from the compression-expansion cycles emphasize the important contribution of the terminal n-octadecyl chain to the stability of the polymeric interfacial layer under conditions of high compression, even though it contributes only weakly to the initial surface activity of PNIPAM derivatives, which is controlled primarily by the NIPAM units. Similar observations have been reported in a comparative study of poly(ethylene oxide) (PEO) and its telechelic derivatives carrying two terminal n-hexadecyl chains (36-39).

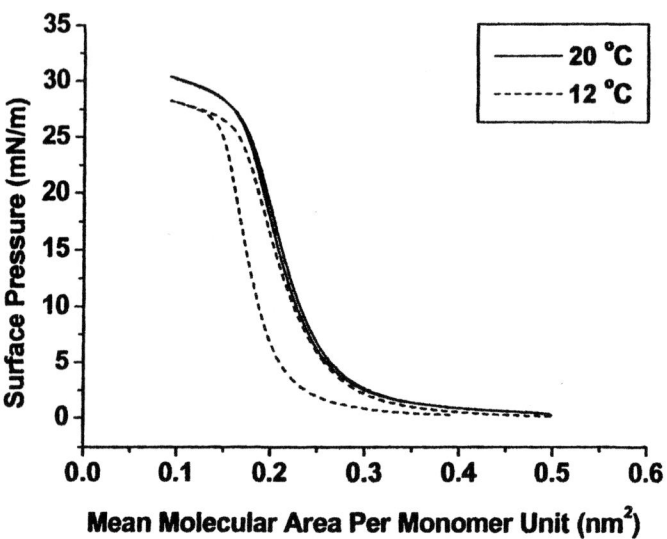

Figure 6. Compression-expansion cycles of the monolayers of C_{18}-PNIPAM-C_{18} at two subphase temperatures.

While the overall properties of PNIPAM derivatives at the air/water interface are quite similar, they differ in a number of subtle features, which are diagnostic of distinct changes in the conformation of the various polymers as a function of their molecular architecture, as depicted in Figures 7 and 8.

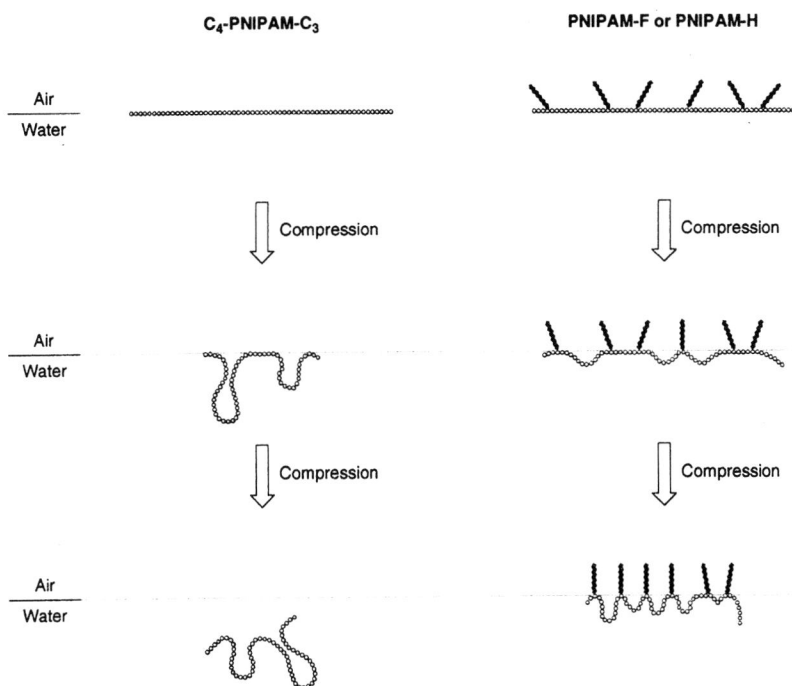

Figure 7. Conceptual representations of the graft PNIPAM derivatives at the air/water interface.

Significant differences are observed under high compression conditions. In the case of C_4-PNIPAM-C_3 the surface pressure eventually reaches a constant value. In highly compressed monolayers, the water soluble C_4-PNIPAM-C_3 tends to escape the air/water interface and dissolve in the aqueous subphase to avoid increasingly strong repulsive interactions among polymer chains. As a result, the polymer surface concentration at the air/water interface remains constant. Monolayers of C_{18}-PNIPAM-C_3 and C_{18}-PNIPAM-C_{18}, however, do not follow the same pattern: their isotherms are characterized by a gentle but *continuous* increase in surface pressure. Introducing a terminal *n*-octadecyl chain, which behaves as an anchor of the PNIPAM main chain at the interface, increases the stability of the corresponding (C_{18}-PNIPAM-C_3) monolayer. In the case of telechelic PNIPAM derivative (C_{18}-PNIPAM-C_{18}), the two terminal *n*-octadecyl chains act in concert, further shifting the hydrophobic-hydrophilic balance. The anchored *n*-octadecyl chains force the PNIPAM main chains to remain close to the water surface, where they are subjected to interchain repulsive forces counterbalanced by the hydrophobic interactions among the hydrophobic tails.

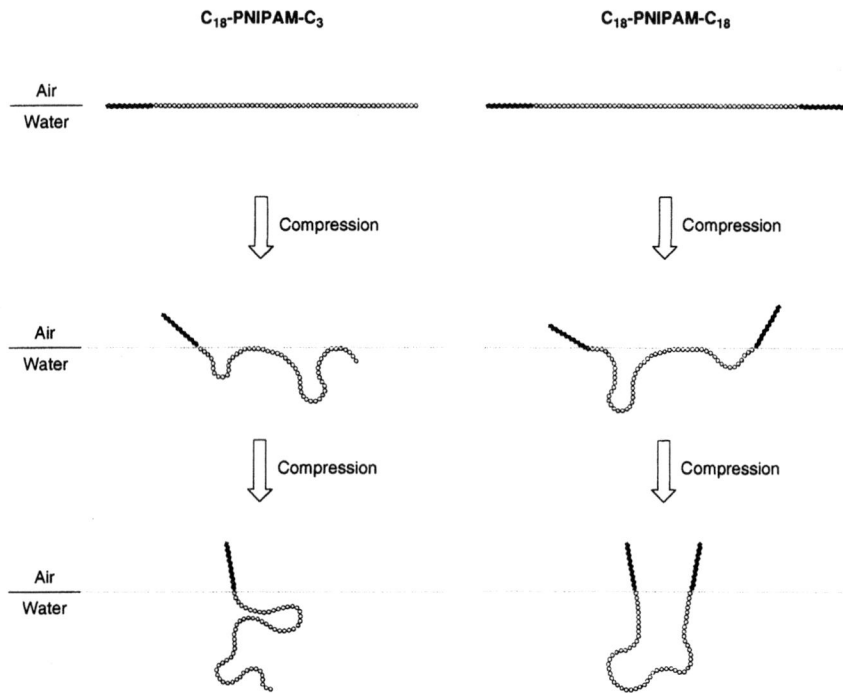

Figure 8. Conceptual representations of the end-functionalized PNIPAM derivatives at the air/water interface.

The effect of temperature on compression isotherms may also be understood based on this analysis. The surface activity of the unmodified PNIPAM (C_4-PNIPAM-C_3) increases with increasing subphase temperature, from 12 to 20 °C. In fact, C_4-PNIPAM-C_3 becomes even more surface active as the subphase temperature reaches ~ 32 °C (data not shown), a tendency characteristic of all the PNIPAM derivatives studied here. This trend is consistent as well with earlier reports of experiments carried out with a PNIPAM sample of high molecular weight (600 000 Da) and broad molecular weight distribution (3.23) (*17*). Given the low level of modification of the various polymers (the NIPAM content of the end-functionalized PNIPAM derivatives is ≥ 99 mol% and that of the graft-PNIPAM derivatives is 90 mol%), it is not surprising that the interfacial properties are dominated by those of the NIPAM segments (*40*). Further studies on the interfacial properties of various PNIPAM derivatives are currently in progress, in order to examine the effects of polymer molecular weight and degree of hydrophobic modification on their interfacial properties, as the subphase temperature nears and exceeds the phase transition temperature of the respective polymers in bulk solutions.

Summary

The stability of PNIPAM Langmuir films can be enhanced significantly via hydrophobic modification of the polymer, whether on the chain ends or randomly along the chain. The stabilizing effect of hydrophobic substituents is notable, even keeping the hydrophobic group content as low as 10 mol%, in particular in the case of fluorinated n-alkyl grafts. The interfacial properties and stability of the monolayers vary as a function of the subphase temperature for temperatures well below the cloud point of aqueous PNIPAM solutions. The Langmuir film balance technique provides a method to study the temperature-responsiveness of PNIPAM chain fragments of hydrophobic PNIPAM derivatives that are insoluble in water and cannot be examined by solution-based experimental techniques.

Acknowledgments

The work was supported in part by the Natural Sciences and Engineering Research Council of Canada and in part by the American Chemical Society Petroleum Research Fund (grant # 36656-AC7). The authors acknowledge the contributions of Charbel Diab, Julie Murray, and Sherry Shaban to part of this work.

References

1. Schild, H. G. *Prog. Polym. Sci.* **1992**, *17*, 163.
2. Spafford, M.; Polozova, A.; Winnik, F. M. *Macromolecules* **1998**, *31*, 7099.
3. Principi, T.; Goh, C. C. E.; Liu, R. C. W.; Winnik, F. M. *Macromolecules* **2000**, *33*, 2958.
4. Kujawa, P.; Goh, C. C. E.; Calvet, D.; Winnik, F. M. *Macromolecules* **2001**, *34*, 6387.
5. Kujawa, P.; Liu, R. C. W.; Winnik, F. M. *J. Phys. Chem. B* **2002**, *106*, 5578.

6. Winnik, F. M.; Winnik, M. A.; Ringsdorf, H; Venzmer, J. *J. Phys. Chem.* **1991**, *95*, 2583.
7. Barros, T. C.; Adronov, A.; Winnik, F. M.; Bohne, C. *Langmuir* **1997**, *13*, 6089.
8. Kujawa, P.; Watanabe, H.; Tanaka, F.; Winnik, F. M. *Eur. Phys. J. E-Soft Matter.* Submitted, 2004.
9. Lee, L. T.; Jean, B.; Menelle, A. *Langmuir* **1999**, *15*, 3267.
10. Jean, B.; Lee, L. T.; Cabane, B. *Langmuir* **1999**, *15*, 7585.
11. Richardson, R. M.; Pelton, R.; Cosgrove, T.; Zhang, J. *Macromolecules* **2000**, *33*, 6269.
12. Huang, Q. R.; Wang, C. H. *Langmuir* **1999**, *15*, 634.
13. Zhang, J.; Pelton, R. *Langmuir* **1996**, *12*, 2611.
14. Noskov, B. A.; Akentiev, A. V.; Bilibin, A. Yu.; Grigoriev, D. O.; Loglio, G.; Zorin, I. M.; Miller, R. *Langmuir* **2004**, *20*, 9669.
15. Fujishige, S.; Koiwai, K.; Kubota, K.; Ando, I. Kenkyu Hokoku-seni Kobunshi Zairyo Konkyusho **1991**, *167*, 47.
16. Zhang, J.; Pelton, R. *Langmuir* **1999**, *15*, 5662.
17. Kawaguchi, M.; Saito, W.; Kato, T. *Macromolecules* **1994**, *27*, 5882.
18. Saito, W.; Mori, O.; Ikeo, Y.; Kawaguchi, M.; Imae, T.; Kato, T. *Macromolecules* **1995**, *28*, 7945.
19. Saito, W.; Kawaguchi, M.; Kato, T.; Imae, T. *Langmuir* **1996**, *12*, 5947.
20. Miyashita, T. ; Konno, M.; Matsuda, M.; Saito, S. *Macromolecules* **1990**, *23*, 3531.
21. Miyashita, T. ; Mizuta, Y.; Matsuda, M. *Br. Polym. J.* **1990**, *22*, 327.
22. Qian, P.; Nanjo, H.; Sanada, N.; Yokoyama, T.; Itabashi, O.; Hayashi, H.; Miyashita, T.; Suzuki, T. M. *Thin Solid Films* **1999**, *349*, 250.
23. Feng, F.; Miyashita, T. *Chem. Lett.* **1999**, 669.
24. Budianto, Y.; Aoki, A.; Miyashita, T. *Macromolecules* **2003**, *36*, 8761.
25. Feng, F.; Aoki, A.; Miyashita, T. *Chem. Lett.* **1998**, 205.
26. Feng, F.; Aoki, A.; Miyashita, T. *Polym. Mater. Sci. Eng.* **1999**, *80*, 413.
27. Li, X.-D.; Aoki, A.; Miyashita, T. *Langmuir* **1996**, *12*, 5444.
28. Shi, X.; Li, J.; Sun, C.; Wu, S. *Colloids Surf. A: Physicochem. Eng. Aspects* **2000**, *175*, 41.
29. Liu, R. C. W.; Segui, F.; Viitala, T.; Winnik, F. M. *Polym. Mater. Sci. Eng.* **2004**, *90*, 105.
30. Moad, G.; Mayadunne, R. T. A.; Rizzardo, E.; Skidmore, M.; Thang, S. H. *Macromol. Symp.* **2003**, *192*, 1.
31. In fact, compressing the monolayers at a rate different from 20 mm/min gave isotherms with different profiles, indicating the slow dynamic nature of the polymer response to compression. Similar phenomena are observed on expansion isotherms. The effect of compression and expansion rate on the isotherm profile will be discussed in detail in an upcoming article.

32. Valli, L. In *Thin Solid Films: Application, Preparation and Characterization, 2003*; Trusso, S.; Mondio, G.; Neri, F., Eds.; Research Signpost: Trivandrum, India 2003; p 71.
33. Kawaguchi, M.; Hirose, Y.; Kato, T. *Langmuir* **1996**, *12*, 3523.
34. Ravey, J. C.; Stebe, M. J. *Colloids Surf. A: Physicochem. Eng. Aspects* **1994**, *84*, 11.
35. Petit, F.; Iliopoulos, I.; Audebert, R.; Szönyi, S. *Langmuir* **1997**, *13*, 4229.
36. Cao, B. H.; Kim, M. W. *Faraday Discuss.* **1994**, *98*, 245.
37. Barentin, C.; Muller, P.; Joanny, J. F. *Macromolecules* **1998**, *31*, 2198.
38. Barentin, C.; Joanny, J. F. *Langmuir* **1999**, *15*, 1802.
39. Renan, M.; François, J.; Marion, D.; Axelos, M. A. V.; Douliez, J.-P. *Colloids Surf. B: Biointerfaces* **2003**, *32*, 213.
40. The opposite trend has been reported in a study of random copolymer of NIPAM (98.5 mol%) and *n*-octadecylacrylate (1.5 mol%). See reference 29. Since the sample has a low polymer molecular weight (M_w = 5198 and M_n = 2912), the sample is essentially composed of PNIPAM and HM-PNIPAM bearing one *n*-octadecyl chain randomly on the polymer backbone.

Chapter 8

Stimuli-Responsive Behavior of Sodium Dodecyl Sulfate and Sodium Dioctyl Sulfosuccinate during Coalescence of Colloidal Particles

David J. Lestage, W. Reid Dreher, and Marek W. Urban*

School of Polymers and High Performance Materials, Shelby F. Thames Polymer Science Research Center, Department of Polymer Science, The University of Southern Mississippi, Hattiesburg, MS 39406

Over the last decade significant advances have been made in developing further understanding of how minute chemical and physical changes in multi-component colloidal dispersions may have significant impact on many microscopic film properties. This chapter focuses on stimuli-responsive behavior of two commonly utilized surfactants, sodium dodecyl sulfate (SDS), and sodium dioctyl sulfosuccinate (SDOSS) in the synthesis of colloidal dispersions as well as a result of film formation under various conditions. The effects of SDS and SDOSS structural similarities and differences are discussed in the context of external and internal stimuli such as temperature, pH, ionic strength, particle morphology, and covalent and ionic crosslinkers present in an aqueous phase, and the resulting chemical and physical properties during and after coalescence.

Introduction

Film formation from colloidal dispersions has been and continues to be of significant interest. The current understanding of colloidal film formation dates back almost half a century where interparticle contact, driven by capillary forces of evaporating water[1-3] and subsequent initiation of coalescence involving diffusion of polymer chains across particle boundaries formulated the basis for other studies which, in many cases, are repetitive efforts of the earlier proposed phenomenon.[4,5] In spite of often simplified views on coalescence, many outstanding research papers and monographs were published which clearly indicated that colloidal systems are sensitive to minute chemical and physical changes. As a consequence, film formation processes proposed in the past have achieved conceptual and scholastic importance, but for those with experience in synthesis and preparation of colloidal dispersions agree there are many experimental variables and considerations that may lead to significant discrepancies. It also became clear that altering macromolecular chains of colloidal particles by, for example, fluorescing agents or other species that may track the macromolecular mobility, also alter processes responsible for coalescence. Although one could argue that such small quantities should not influence coalescence conditions, usually minute alterations are responsible for major macroscopic changes. To further advance our knowledge on coalescence, advanced techniques such as scanning and transmission electron or atomic force microscopies have provided visualization of particle packing, and based on these studies, lattice packing and particle deformation was established along with the affects of colloidal particle size and processes governing crystallite formation during annealing of coalesced films.[6]

Over the last decade, spectroscopic techniques providing molecular level information[7-9] focused primarily on mobility[10-13] of low mol. wt. species, in particular surfactant molecules. These studies have shown that such intrinsic polymer properties as glass transition temperature, the chemical makeup of colloidal particles and external coalescence conditions may also play an essential role on coalescence. For example, temperature and relative humidity (RH) are of the main external stimuli, but less recognizable are surface tensions as well as interactions of individual components.[14,15] In an effort to recognize the importance of stimuli-responsive behavior of colloidal dispersions, numerous studies[10,16-38] have been conducted on stratification and mobility of individual components in coalesced polymer films obtained from colloidal dispersions. Controlled mobility of such surfactants as well as other low mol wt. species may impart stimuli-responsive behaviors, and distribution of these species across coalesced films are greatly effected by temperature, pH, ionic strength, and other factors.[10,14,20,39-49] This chapter focuses on the nature of these interactions and how film interfacial structures are affected.

Interfacial Interactions

Ionic interactions between sodium dioctyl sulfosuccinate (SDOSS) and ethyl acrylate/methacrylic acid (EA/MAA) copolymer dispersions were characterized at both the film-air (F-A) and film-substrate (F-S) interfaces using attenuated total reflectance Fourier transform infrared spectroscopy (ATR-FTIR),[37,38] and these studies showed that upon film formation, SDOSS is present at the F-A interface and its hydrophilic end reacts with H_2O as well as MAA components. This is schematically illustrated in Figure 1 which depicts coalesced SDOSS surface stabilized particles under dry conditions (A) followed by interactions of surface species with the continuous phase (B) and SDOSS association with acid species in the polymer matrix (C). Upon the neutralization of MAA, the presence of the 1056 cm^{-1} IR band diminished thus giving rise to only $SO_3^-Na^+$---H_2O interactions at the F-A interface (1046 cm^{-1}).

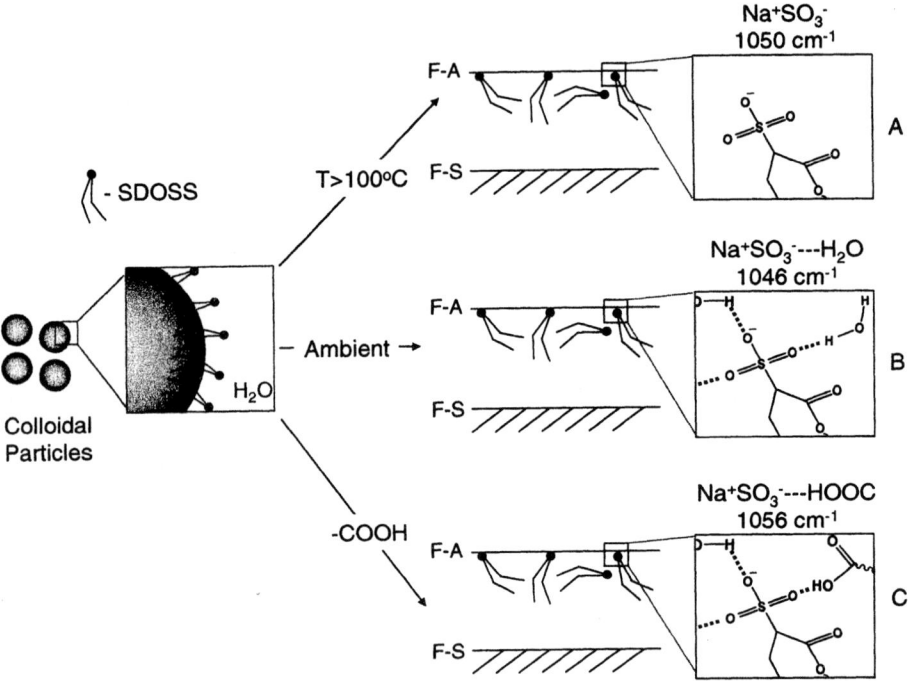

Figure 1. Schematic illustration vibrational dependencies of SDOSS: A – high temperature and low humidity, B – interactions with aqueous environment, C – association with acid species present in polymer matrix.

Other studies[31-33] have shown that $SO_3^-Na^+$---HOOC interactions may be displaced in the presence of an appropriate stimuli, such as increased relative humidity (RH) during coalescence, resulting in a reduced compatibility between SDOSS and EA/MAA which gives rise to an increased concentration of SDOSS near the F-A interface.

Migration of SDS to the F-A or F-S interfaces can also be affected by the chemical makeup up colloidal particles. For example, in the presence of ethylhexyl acrylate/methacrylic acid (EHA/MAA) colloidal particles,[50] the F-S interfacial concentration of SDS is directly related to the surface tension of the substrate utilized during coalescence. When polytetrafluoroethylene (PTFE) was used as a substrate, SDS migrated to both the F-A and F-S interfaces, but due to significant interfacial surface tension difference between the substrate and the aqueous phase, SDS migration to the F-S interface is dominant. By changing the substrate to glass, subsequently decreasing the interfacial surface tension difference, a decrease at the F-S interface of SDS in EHA/MAA colloidal films occurs.

Crosslinking and Particle Morphology

Another avenue of particular interest is the influence of chemical stimuli such as the presence of crosslinking agents on the mobility of surfactant molecules. It is well known that the presence of crosslinkable moieties reinforces a polymer network, which may alter mobility of species that are not chemically bound within a polymer matrix. As a result, surface morphology and orientation of surface species at interfacial regions may be altered. Recent studies on styrene/ethyl hexyl acrylate/methacraylic acid (Sty/EHA/MAA) colloidal films[15] have shown that upon incorporating adipic dihydrazide (ADDH) and diacetone acrylamide (DAAM) as covalently binding crosslinking agents, coalescence of Sty/EHA/MAA was inhibited. This, in turn, gave rise to a decrease in the surface concentration of SDS, but elevated temperatures caused stratification of SDS near the F-A interface. Similarly, ionic strength of the solution may serve as another stimuli. For example, when $Ca(OH)_2$ and NH_4OH are utilized as ionic crosslinkers, the presence of the Ca^{2+} ion does not inhibit particle coalescence which results in the migration of SDS to the F-A interface under ambient conditions. Furthermore, Ca^{2+} ion is able to complex with the hydrophilic head group of SDS and quantum mechanical calculations showed that hydrophilic head groups change conformations depending upon the presence of different counter ions. These changes are illustrated in Figure 2. Similarly, orientation of SDOSS and SDS in styrene/n-butyl acrylate/methacrylic acid (Sty/nBA/MAA) colloidal films is such that hydrophobic chains are preferentially perpendicular to the surface and their hydrophilic ends ($-SO_3^-Na^+$) associated with H_2O and $-COOH$ entities tend to align parallel to the film surface,[13] and the S-O bond of the $-SO_3^-Na^+$ group of

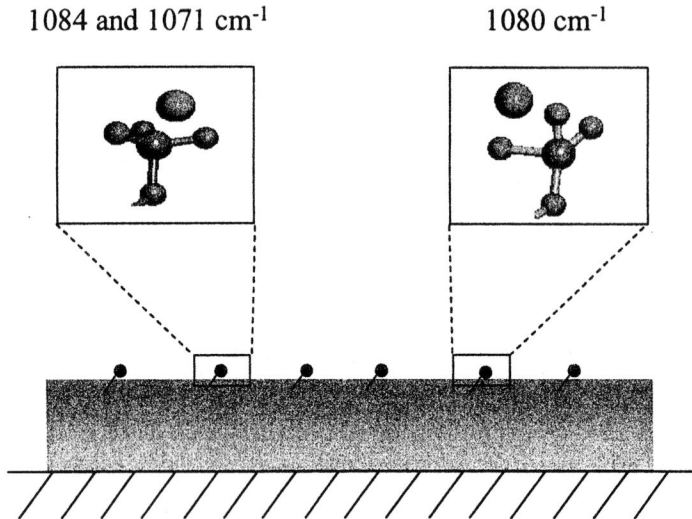

Figure 2. Preferential orientation of $-SO_3^-Na^+$ groups near the surface of Sty/EHA/MAA (Adapted from reference 15. Copyright 2003 American Chemical Society.) (See page 2 of color insert.)

SDOSS or SDS can take either out-of-plane or in-plane conformations, depending on their environments.

There is also significant effect of particle morphology on stratification behavior. For example, upon particle coalescence, epoxy and acid functionalities of Sty/nBA/GMA and Sty/nBA/MAA can crosslink, result in added stability to the colloidal film. However, elevated coalescence temperatures cause crosslinking reactions to occur much faster between epoxy and acid groups, thus a decrease in the surface concentration of SDOSS occurs.[14]

Particle composition can significantly alter migrational responses of surfactants during film formation. For example, minor concentrations of SDS were observed at the F-A interface for EHA/MAA colloidal particles, but by altering the copolymer matrix to methyl methacrylate/n-butyl acrylate (MMA/nBA),[39,43,47,51] SDS concentration levels are higher. Thus, SDS migrates to both the F-A and F-S interfaces in MMA/nBA colloidal films, but due to a weak water flux its stratification at the F-S interface is significantly higher than at the F-A interface. Mobility of low mol. wt. components may also be affected by the stability of colloidal dispersions. As was shown,[51,31] when surfactant molecules are displaced from the particle surface as a result of time-dependent limited compatibility, free SDOSS surfactant may stratify near the F-A interface.

Altering colloidal particle morphology[35] to consists of MMA and nBA results in contrastingly different barrier properties which may influence stimuli-responsive behaviors in coalesced films. As shown in Figure 3, exudation of SDOSS carried by H_2O at various stages of film coalescence was investigated in (A) p-MMA, (B) MMA/nBA copolymer, (C) p-MMA/p-nBA core-shell, (D) p-MMA/p-nBA blend, (E) and p-nBA dispersions. These studies have shown that in early stages of coalescence, homo- and copolymer matrices are permeable to water allowing diffusion both into and around colloidal particles. However, distinction occurs due to free volume and hydrophobic properties whereupon particles containing primarily p-MMA limit diffusion through particles, and furthermore, retain absorbed H_2O for longer periods of time. In the final stages of coalescence, primarily p-nBA containing particles reveal a reduced rate of mass loss due to evaporation, however exhibit the largest mass lost.

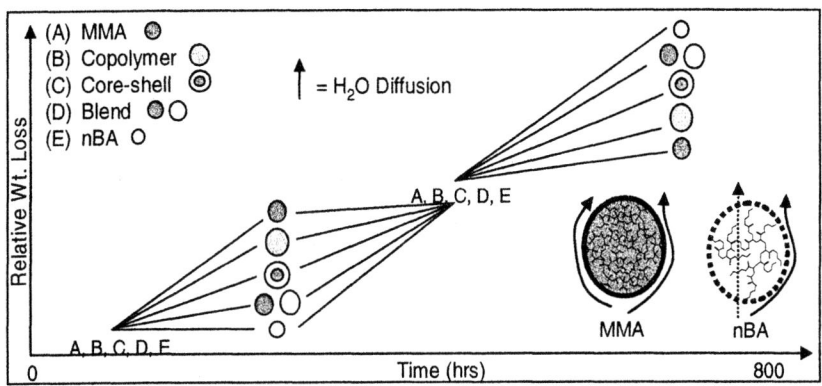

Figure 3. Schematic representation of mobility of SDOSS and H_2O at different stages of film formation (Reproduced from reference 35. Copyright 2004 American Chemical Society.)

Surfactant-Particle Interactions

More recently[10,16-18,35,52] mobility of SDOSS surfactant as a result of ionic interactions in Sty/nBA/MAA colloidal dispersions composed of copolymer, blend, and core-shell particle morphologies was investigated, and a non-uniform distribution of SDOSS across the film thickness was observed, but distribution of surfactant molecules in a colloidal films can be greatly affected by particle morphology, temperature, substrate effects, and alternate ionic environments.[10,14,20,39-49] For example, surfactant stratification in Sty/nBA/MAA colloidal films is directly related to annealing temperature and the degree of neutralization of MAA[20]. Figure 4, a, illustrates that as external stimuli such as temperature above the T_g of the film, results in the increased surface concentration of SDOSS.

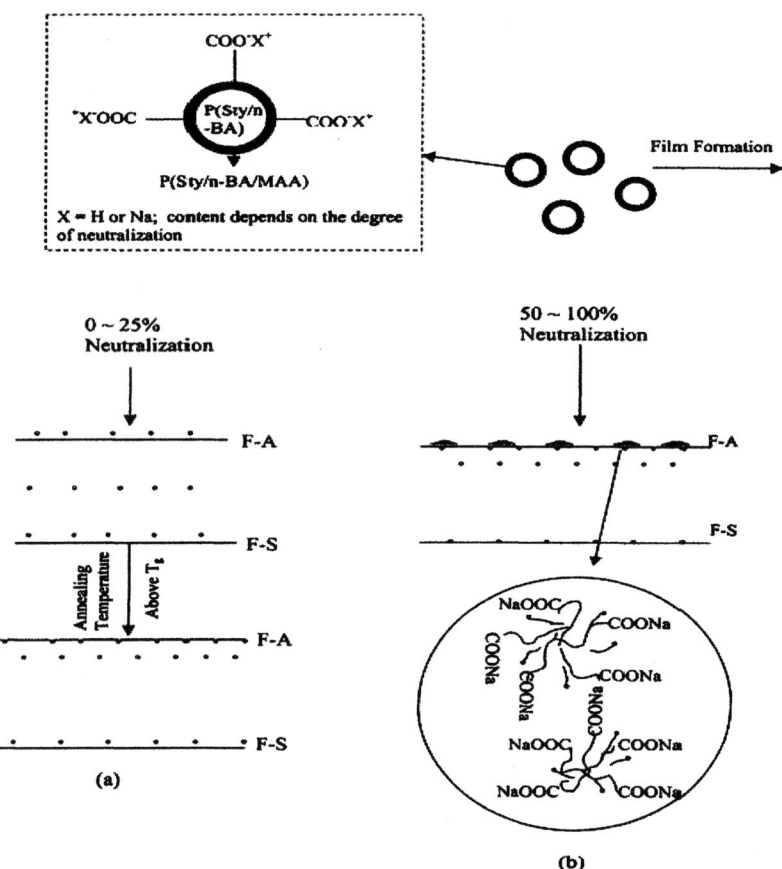

Figure 4. Schematic representation of the film formation process of Sty/nBA core and MAA shell colloidal particles: a – F-A and F-S distribution of SDOSS as a function of annealing temperature. b – Ionic interactions present between MAA and SDOSS disrupted via 50 – 100% neutralization and subsequent release of SDOSS to the F-A interface (Adapted from references 19 and 20. Copyright 2001 and 2002 American Chemical Society.)

Furthermore, as shown in Figure 4, b, upon neutralizing MAA (50-100%), SDOSS migrates to the F-A interface in the presence of ionic clusters which supply added physical properties to the colloidal film, but when MAA was neutralized to a lesser extent (0-25%), SDOSS remained suspended throughout the film without a significant accumulation at the F-A interface.

Although one may argue that stimuli-responsive stratification of low molecular weight species in colloidal films may be undesirable, recent studies[53,54] have shown that interactions between surfactants and polymers may be advantageous in previously unobtainable advanced applications. Numerous investigations[49,54-64] focused on interactions between the copolymer matrix, alternate ionic and nonionic surfactants, as well as electrolyte solutions. One example is how hydrophobic tails of SDS can be greatly effected by Sty/AA[55] and MMA/EHA/MAA[49] matrices. Replacement of AA with MAA resulted in limited migration of SDS to the F-A interface due to SDS and the polymer matrix interactions, and also altered interactions between the acid functionalities of the copolymer and the hydrophilic head group of the surfactant molecules. Furthermore, the presence of various counterions and alternate hydrated forms of SDS, crystallites of different sizes and shapes can be obtained[57] which can significantly alter film formation processes.

Local Ionic Clusters

Recent studies[65] also addressed the formation of ionic domains which inhibit the exudation of SDS from the bulk of the film to the F-A interface during film formation. For example, in the presence of MMA/nBA/AA, no SDS was observed at the F-A interface upon particle coalescence, but when MMA/nBA was utilized, SDS was able to stratify at the surface upon film formation. As shown in Figure 5, a, interactions between the ester functionality of the polymer backbone and $SO_3^-Na^+$ groups occur through the Na^+ ion which can be easily mobilized in the presence of H_2O, but the presence of AA increases the strength of the $SO_3^-Na^+\cdots COOH$ interactions, thus introducing more symmetry to the $SO_3^-Na^+$ environment. This gives rise to the formation of localized ionic clusters (LICs), which facilitate an environment that exhibits significantly higher thermal stability as compared to MMA/nBA colloidal dispersions. Incorporation of propylene glycol (PG) generates desirable responses in the presence of MMA/nBA/AA allowing for the interactions between AA and SDS to be disrupted by the solubilization of SDS from the surface of MMA/nBA/AA particles. This is shown in Figures 5, b and c, where PG serves as a vehicle for the deposition of SDS at the F-A interface. The SDS migration results in the formation of heterogeneous islands of an alternate

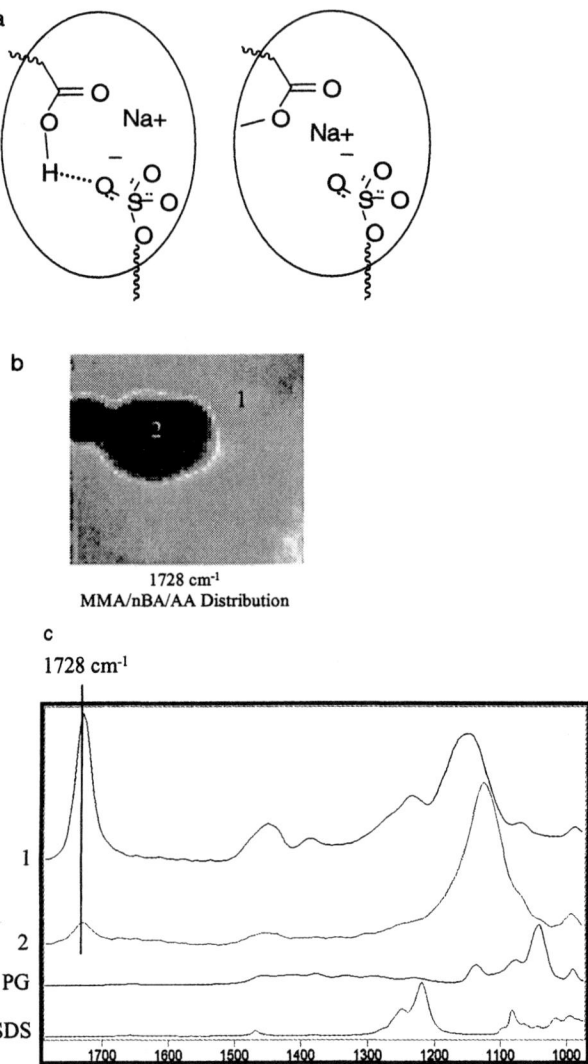

Figure 5. a – Interactions between $-SO_3^-Na^+$ groups of SDS and acid groups in the presence of MMA/nBA and MMA/nBA/AA copolymers; b – IRIRI image obtained by tuning into the 1728 cm^{-1} IR band after 24 hours of coalescence. (See page 3 of color insert.) c – Averaged IR spectra recorded from regions 1 and 2 as labeled in Figure 5, b (Reproduced from reference 65. Copyright 2003 American Chemical Society.)

crystalline form as shown in the IRIRI image in Figure 5, b, and the IR spectrum of the area labeled Region 2 in Figure 5, c.

Similar entities may be generated as a result of molecular interactions between SDOSS and hydrogenated soybean phosphatidylcholine (HSPC) phospholipid in the presence of ionic environments.[34] When an SDOSS/HSPC mixture is utilized to stabilize p-MMA/nBA copolymer dispersion, stimuli-responsive crystal formation occurs when exposed to various ionic strength $CaCl_2$ solutions. This is illustrated in Figure 6, where three stages of molecular interactions during the critical stages of coalescence are illustrated. As seen, coalesced films containing no electrolyte solutions exhibit homogeneous surfaces, as determined by IRIRI tuned to the symmetric S-O stretching of SDOSS at 1050 cm^{-1}, as a result of ionic binding of SDOSS and HSPC polar moieties.

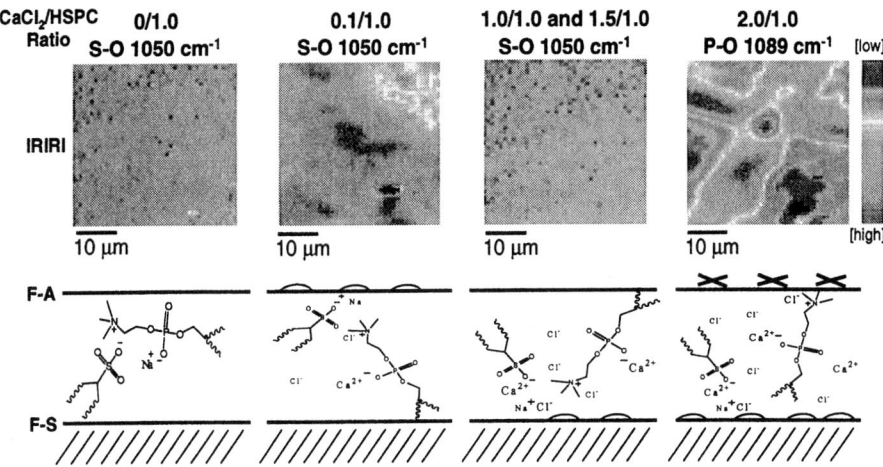

Figure 6. Schematic representation of particle interactions: A – without Ca^{2+}; B and C – in the presence of Ca^{2+} ions and the effect of $CaCl_2$/HSPC ratio on mobility and crystallization at the F-A interface (Reproduced from reference 34. Copyright 2004 American Chemical Society.) (See page 3 of color insert.)

Upon addition of low $CaCl_2$/HSPC molar ratios, localized ionic interactions within the polymer matrix between SDOSS and HSPC hydrophilic ends are disrupted, thus allowing the release of SDOSS to the F-A interface. However, at higher $CaCl_2$/HSPC molar ratios of 2.0/1.0, release and subsequent crystallization of HSPC domains occurs to the F-A interface, and when tuned to symmetric P-O stretching of HSPC at 1089 cm^{-1}, IRIR images reveal star-like crystals at the F-A interface with enhanced concentrations of HSPC.

Interactions in the Presence of Fluoro-Surfactants

As shown in previous studies,[10,40] stratification near interfaces may be driven by a number of factors, and the chemical makeup along with the surface energy of the surface stabilizing species can play a significant role on the surface self-assembly of SDS in the presence of other surfactants. Thus, chemical structures of surfactants in a given environment will dictate their ability to alter particle interfacial properties. For example, incorporation of F-containing surfactants may provide desirable surface properties. Recent studies[66] utilized MMA/nBA colloidal dispersions which were synthesized in the presence of SDS and incorporated F-containing surfactants of different ionicity and chemical structures. As a result, different morphological features were obtained and are shown in Figure 7, a-d. Figure 7, a, represents featureless surfaces which were generated when no F-containing surfactant was utilized. However, the incorporation of anionic F-containing surfactants (Figure 7, b and d), results in mixed micellar domains formed as heterogeneous spheres consisting of the F-containing surfactant and SDS. In the presence of a non-ionic F-containing surfactant, (Figure 7, c), intermolecular cohesion present in the F-containing surfactant results in a rod-like morphology resulting from incompatibility between the surfactants. Furthermore, the chemical makeup of the F-containing surfactant greatly affects the mobility of SDS. This is illustrated in Figure 7, e which shows that SDS concentration levels are significantly altered near the F-A interface and its magnitude depends on the chemical structure of F-containing surfactant.

Conclusions

The film formation of colloidal dispersions from the continuous aqueous phase is a complex process where interactions between individual components in the aqueous phase and non-continuous phase must be considered. Incorporation of chemical and physical stimuli into these multi-component systems may result in desirable or undesirable responses which can be achieved by chemico-physical design of colloidal dispersion and processes responsible for film formation. As was shown above, responses of SDS and SDOSS can be greatly affected by copolymer microstructures, particle morphology, acidic/basic environments, temperature, and covalent and ionic crosslinking moieties, as well as ionic strength. However, there are other factors that need to be further investigated in order to fully utilize stimuli-responsive behaviors of this technologically important family of water-dispersed polymers.

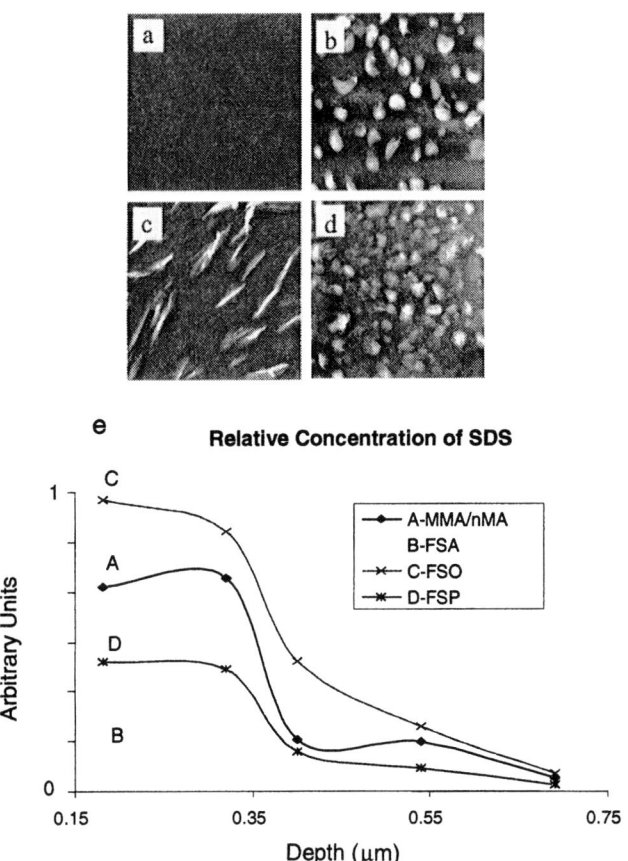

Figure 7. a – ESEM micrographs of the F-A interface of MMA/nBA film: A – without fluorocarbon surfactant; B – FSA; C – FSO; and D – FSP. b – Relative concentration levels of SDS plotted as a function of depth from the F-A interface for: A – MMA/nBA containing FSA; B – MMA/nBA containing FSO; and C – MMA/nBA containing FSP (see page 4 of color insert.) (Reproduced from reference 66. Copyright 2004 American Chemical Society.)

Acknowledgments

The authors thank numerous industrial sponsors and National Science Foundation for partial support (DMR 0213883) of these studies.

References

1. Vanderhoff, J. W.; Tarkowski, H. L.; Jenkins, M. C.; Bradford, E. B. *Jounal of Macromolecular Chemistry* **1966**, *1*, 131.
2. Brown, G. L. *J. Polym. Sci.* **1956**, *22*, 423.
3. Voyutskii, S. S. *J. Polym. Sci.* **1958**, *32*, 528.
4. Wang, Y.; Zhao, C.; Winnik, M. *J. of Chem. Phys.* **1991**, *95*, 2143.
5. Anderson, J.; Jou, J. *Macromolecules* **1987**, *20*, 1544.
6. Wang, Y.; Kats, A.; Juhue, D.; Winnik, M.; Shivers, R.; Dinsdale, C. *Langmuir* **1992**, *8*, 1435.
7. Oh, J.; Tomba, P.; Ye, X.; ELey, R.; Rademacher, J.; Farwaha, R.; M., W. *Macromolecules* **2003**, *36*, 5804.
8. Goudy, A.; Gee, M.; Biggs, S.; Underwook, S. *Langmuir* **1995**, *11*, 4454.
9. van Tent, A.; te Nijenhuis, K. *J. Coll. Int. Sci.* **2000**, *232*, 350.
10. Zhao, Y.; Urban, M. W. *Macromolecules* **2000**, *33*, 2184.
11. Niu, B. J.; Urban, M. W. *J.Appl.Pol. Sci.* **1996**, *60*, 371.
12. Niu, B. J.; Urban, M. W. *Film Formation in Waterborne Coatings*; American Chemical Society: Washington, DC, 1996.
13. Zhao, Y.; Urban, M. W. *Langmuir* **2001**, *17*, 6961.
14. Zhao, Y.; Urban, M. *Macromolecules* **2000**, *33*, 8426.
15. Dreher, W. R.; Zhang, P.; Urban, M. W.; Porzio, R. S.; Zhao, C. *Macromolecules* **2003**, *36*, 1228.
16. Zhao, Y.; Urban, M. W. *Langmuir* **2000**, *16*, 9439-9447.
17. Zhao, Y.; Urban, M. W. *Macromolecules* **2000**, *33*, 7573-7581.
18. Zhao, Y.; Urban, M. W. *Macromolecules* **2000**, *33*, 8426-8434.
19. Zhao, Y.; Urban, M. W. *Langmuir* **2001**, *17*, 6961-6967.
20. Zhao, Y.; Urban, M. W. *Langmuir* **2000**, *16*, 9439.
21. Niu, B. J.; Urban, M. W. *J. Appl. Polym. Sci.* **1996**, *62*, 1903-1911.
22. Niu, B. J.; Urban, M. W. *J. Appl. Polym. Sci.* **1996**, *60*, 371-377.
23. Niu, B. J.; Urban, M. W. *J. Appl. Polym. Sci.* **1996**, *60*, 379-387.
24. Niu, B. J.; Urban, M. W. *J. Appl. Polym. Sci.* **1996**, *62*, 1903-1911.
25. Niu, B. J.; Urban, M. W. *J. Appl. Polym. Sci.* **1998**, *70*, 1321-1348.
26. Keddie, J. L. *Mater. Sci. and Eng.* **1997**, *21*, 101-170.
27. Beltran, C. M.; Guillot, S.; Langevin, D. *Macromolecules* **2003**, *36*, 8506-8512.

28. Sethumadhavan, G. N.; Nikolov, A.; Wasan, D. *Langmuir* **2001**, *17*, 2059-2062.
29. Shin, J. S.; Lee, D. Y.; Ho, C. C.; Kim, J. H. *Langmuir* **2000**, *16*, 1882-1888.
30. Tebelius, L. K.; Urban, M. W. *J. Appl. Polym. Sci.* **1995**, *56*, 387-395.
31. Thorstenson, T. A.; Tebelius, L. K.; Urban, M. W. *J. Appl. Polym. Sci.* **1993**, *50*, 1207-1215.
32. Thorstenson, T. A.; Tebelius, L. K.; Urban, M. W. *J. Appl. Polym. Sci.* **1993**, *49*, 103-110.
33. Thorstenson, T. A.; Urban, M. W. *J. Appl. Polym. Sci.* **1993**, *47*, 1381.
34. Lestage, D. J.; Urban, M. W. *Langmuir* **2004**, *20*, 7027-7035.
35. Lestage, D. J.; Urban, M. W. *Langmuir* **2004**, *20*, 6443.
36. Kunkel, J.; Urban, M. W. *J. Appl. Polym. Sci.* **1993**, *50*, 1217-1223.
37. Evanson, W.; Thorstenson, T. A.; Urban, M. W. *J. Appl. Polym. Sci.* **1991**, *42*, 2297-2307.
38. Evanson, W.; Urban, M. W. *J. Appl. Polym. Sci.* **1991**, *42*, 2287-2296.
39. Zhao, C. L.; Holl, Y.; Pith, T.; Lambla, M. *Coll. Polym. Sci.* **1987**, *265*, 823.
40. Zhao, Y.; Urban, M. *Macromolecules* **2000**, *33*, 8426.
41. Zhao, Y.; Urban, M. *Macromolecules* **2000**, *33*, 7573.
42. Zhao, Y.; Urban, M. *Langmuir* **2001**, *17*, 6961.
43. Holl, Y. *Macrom. Symp.* **2000**, *151*, 473.
44. Reiter, G.; Khanna, R.; Sharma, A. *J. Phys.: Cons. Matt.* **2003**, *15*, S331.
45. Mallegol, J.; Gorce, J.; Dupont, O.; Jeynes, C.; McDonald, P.; Keddie, J. *Langmuir* **2002**, *18*, 4478.
46. Aydogan, N.; Abbott, M. *Langmuir* **2001**, *18*, 5703.
47. Keitz, E.; Holl, Y. *Colloids and Surfaces A: Physiochemical and Engineering Aspects* **1993**, *78*, 255.
48. Lam, S.; Hellgren, A.; Sjoberg, M.; Holmberg, K.; Schoonbrood, H.; Unzue, M.; Asua, J.; Tauer, K.; Sherrington, D.; Montoya Goni, A. *J. Appl. Polym. Sci.* **1997**, *66*, 187.
49. Amalvy, J.; Soria, D. *Prog. in Org. Coat.* **1996**, *28*, 279.
50. Evanson, K.; Urban, M. *J. Appl. Polym. Sci.* **1991**, *42*, 2309.
51. Zhao, C.; Dobler, F.; Pith, T.; Holl, Y.; Lambla, M. *J. Coll. Int. Sci.* **1989**, *128*, 437.
52. Zhao, Y.; Breitenkamp, K.; Urban, M. W. *Polym. Mater. Sci. Eng.* **2000**, *82*, 380-381.
53. Barreiro-Iglesias, R.; Alvarez-Lorenco, C.; Concheiro, A. *J. Contol. Rel.* **2003**, *93*, 319.
54. Lee, J.; Moroi, Y. *Langmuir* **2004**, *20*, 4376.
55. Urban, M.; Koenig, J. *Appl. Spec.* **1987**, *41*, 1028.
56. Wei, H.; Shen, Q.; Zhao, Y.; Wang, D.; Xu, D. *J. Cryst. Grow.* **2004**, *260*, 545.
57. Sperline, R. *Langmuir* **1997**, *13*, 3715.

58. Smith, L. A.; Hammond, R. B.; Roberts, K. J.; Machin, D.; McLeod, G. *J. Mol. Structure* **2000**, *554*, 173.
59. Sehgal, P.; Doe, H.; Bakshi, M. *Coll. Int. Sci.* **2003**, *261*, 584.
60. Robertson, D.; Hellweg, T.; Tiersch, B.; Koetz, J. *Coll. Int. Sci.* **2004**, *270*, 187.
61. Lauten, R.; Kjoniksen, A.; Nystrom, B. *Langmuir* **2001**, *19*, 2947.
62. Khanal, A.; Li, Y.; Takisawa, N.; Kawasai, N.; Oishi, y.; Nakashima, K. *Langmuir* **2004**, *20*, 4809.
63. Xu, Z.; Ducker, W.; Israelachvili, J. *Langmuir* **1996**, *12*, 2263.
64. Meszaros, R.; Thompson, L.; Bos, M.; Varga, I.; Gilanyi, T. *Langmuir* **2003**, *19*, 609.
65. Dreher, W. R.; Urban, M. W.; Porzio, R. S.; Zhao, C. *Langmuir* **2003**, *19*, 10254.
66. Dreher, W. R.; Urban, M. *Langmuir* **2004**, 20, 10455.

Chapter 9

Fluorescence Methods for Latex Film Formation

Önder Pekcan[1] and Ertan Arda[2]

[1]Department of Physics, İstanbul Technical University, Maslak 80626, İstanbul, Turkey
[2]Department of Physics, Trakya University, 22030 Edirne, Turkey

Introduction

In order to be able to form a film by evaporation, latex particles have to be soft so that they can deform under surface and capillary forces and yield to a void free film. Moreover, polymer chains in latex have to be mobile to interdiffuse between adjacent particles to form a homogeneously perfect film. However for many applications such as paints or varnishes, the final film is rigid and water-insensitive. These two important requirements, ability to form a film on one hand and rigidity and water resistance on the other, are opposes to each other and difficult to achieve. In order to overcome this difficulty organic solvents are generally used as the coalescing aids for the latex film formation (*1,2*). They plasticize the latex particles and evaporate after the film formation process has been accomplished. The final product is then rigid and insoluble in water.

Historically, the process of film formation has been divided into three or more stages (*3-6*). Definitions of the stages differ amongst authors, although the generally accepted mechanism consist of: I, Evaporation from a stable dispersion which brings the particles into some form of close packing; II, Deformation of particles leads to a structure without voids, although with the original particles still distinguishable; III, Interdiffusion of polymer chains across particle boundaries yields a continuous film with mechanical integrity and the original particles no longer distinguishable. The historical review of these stages is given in reference *6*.

The developments of techniques of neutron scattering and non-radiative energy transfer have enabled to carry out some experiments to study interdiffusion of polymer molecules across particle-particle boundaries. SANS was employed to measure the extend of interdiffusion in polystyrene (PS) latex having high molecular weight and small particle size, where an increase in the radius of gyration (R_g) of the

© 2005 American Chemical Society

polymer was observed when the system was heated above its glass transition, T_g (7). It was concluded that the polymer chains in the latex particles were reduced in dimension by a factor of four because a large molecule was confined to a small particle consisting of only a few chains. Combination of SANS and tensile strength measurements on PS latex film formation suggest that full mechanical strength was achieved when interpretation was across a distance on the scale of the R_g of the polymer (8). It was concluded that two parameters in determining the rate of interdiffusion in a latex are the location of the chain ends and the ratio of the polymer' s R_g to the radius of the latex particle. Secondary ion mass spectroscopy (SIMS) was used to measured depth profiles at symmetric deuterated-protonated PS interfaces as a function of temperature and molecular weight (9). Non-radiative direct energy transfer (DET) technique was first used to study polymer interdiffusion across particle-particle boundaries (10), which monitors concentration profiles of donor and acceptor dyes that are located initially in separate particles. As interdiffusion occurs mixing of donor and acceptor dyes can be measured by an increase in energy transfer between them. Early measurements on poly(methyl methacrylate) (PMMA) prepared by nonaqueous dispersion polymerization presented that diffusion coefficients are on the order of 10^{-15} cm^2s^{-1} at temperatures between 400-450 °K. The film formation of a poly(butyl methacrylate) (PBMA) latex prepared via emulsion polymerization in water, was studied by the same technique. The diffusion coefficients (10^{-16} cm^2s^{-1}) were determined using spherical diffusion model, and found to be dependent on both time and temperature. A decrease in diffusion rate with time was attributed to the effect of low molecular weight chains near the particle surface dominating at early times. The techniques of data analysis were later developed for melt pressed PMMA particles (11) and latex films (12). It is observed that mass transfer increased as time to a power of ½ as in Fickian diffusion model for low molecular weight polymer. For high molecular weight, there was a distinct ¼ power dependence that cannot be explained by a Fickian or reptation model. Data analysis using DET has improved later by taking into account the donor and acceptor concentration profiles during interdiffusion where a uniform acceptor concentration was considered around a donor in thin slices or shells (13,14). Further a model for DET was developed which considered the heterogeneity in the donor and acceptor concentration profiles (15). Where diffusion coefficient of polymer chain obtained by different DET models and SANS are compared during latex film formation. DET studies in latex blends where one phase is far below its T_g and does not undergo any significant diffusion present that the magnitude of energy transfer is proportional to the interfacial area. Various factors have proven to affect the rate of polymer interdiffusion during latex film formation, which are water (16), coalescing aids (17), organic solvents (18) and cosurfactants (19). Organic solvents which plasticize the latex enhance the rates of polymer interdiffusion. The interdiffusion rate also increase in the presence of cosurfactants. At the times longer than the reptation time, the activation energy for diffusion was found 30 kcal/mol in the presence of cosurfactants, which is 48 kcal/mol in the neat polystyrene. Crosslinking usually does not prevent film formation however it limits the mechanical properties of the final films in certain extent. Besides these several experimental techniques were applied to study latex film formation within the last decade, among which are electron (*20*) and atomic force (*21*) microscopies and steady state (*22*) fluorescence (SSF). Light has been used to monitor film formation for long time. Ellipsometry (*23*) and photon transmission (*24,25*) techniques have been elaborated for latex film formation processes. Scanning (*26*) and transmission (*27*) electron microscopies have been commonly used techniques to study several latex systems during film formation.

In this chapter time resolved, steady state fluorescence and photon transmission techniques will be elaborated for studying latex film formation processes. Organic vapor induced latex film formation from poly(methyl methacrylate) (PMMA) particles will be discussed using fluorescence quenching and direct energy transfer (DET) methods. Molecular weight effect on latex film formation from PMMA particles will be discussed and film formation from nanosized waterborne latexes will be covered. Theoretical models to produce void closure, backbone and healing activation energies will be presented.

Fluorescence Studies

When an organic dye absorbs light, it becomes electronically excited; then fluorescence occurs from the lowest excited singlet state and decays over a time scale of typically nanoseconds (*28*). In addition to unimolecular decay pathways for deexcitation of excited states, a variety of bimolecular interactions can lead to deactivation. These are referred to collectively as quenching processes, which enhance the rate of decay of an excited state intensity, I, as

$$I = A \exp\left(-\frac{t}{\tau}\right) \quad (1)$$

where A is the preexponential factor and τ is the lifetime that characterizes the time scale of excited state decay. For dilute solutions of dye molecules in isotropic media, exponential decays are common. In more complex systems, deviations are often observed. Under these conditions one sometime describes the decay in terms of the mean decay time $\langle \tau \rangle$.

Emission of fluorescence is a radiative transition of an electronically excited molecule from its singlet excited state to its ground state (*28*). Fluorescence quenching normally refers to any bimolecular process between the excited singlet state of a fluorescence dye and a second species that enhances the decay rate of the excited state. One can schematically represent the process as

$$F^* \xrightarrow{k_f, k_{nr}} F \quad (2)$$

$$F^* \xrightarrow{k_q[Q]} F \quad (3)$$

where F and F^* represent the fluorescent molecule and its excited form, and Q is the quencher. k_f, k_{nr} and k_q represent the fluorescence, nonradiative and quenching rate constants, respectively. Many types of processes lead to quenching. Kinetically, quenching processes can be divided into two main categories: dynamic and static. In dynamic quenching, diffusion to form an encounter pair during the excited state lifetime of the dye leads to quenching. Whereas in the static quenching, diffusion does not occur (which is not of interest to us). Dynamic quenching is most likely to occur in fluid solution, where the dye or quencher is free to move. If the quenching rate can be characterized in terms of a single rate coefficient (k_q), and the unquenched decay rate of F, in terms of a unique lifetime, τ_o, then the quenching kinetics will follow the Stern-Volmer equation,

$$\tau^{-1} = \tau_o^{-1} + k_q [Q] \qquad (4)$$

where $[Q]$ represents the quencher concentration.

Vapor-Induced Latex Film Formation

Latex film preparations were carried out in the following manner; the same weights of pyrene (P_y) labeled PMMA particles (29) were dispersed in heptane in a test tube for which solid content was 0.24 %. Film samples were prepared from this dispersion by placing a certain number of drops on a 2.5×0.8 cm^2 glass plates and following the heptane to evaporate. The average film thickness was estimated to be 20 μm. In situ fluorescence decay experiments were performed using Photon Technology International's (PTI) strobe master system (SMS). All the measurements were made at 90° position and slit widths were kept at 20 nm. Experiments were performed in a 1×1 cm^2 quartz cell, which was placed in the SMS, and fluorescence decay was collected over three decades of decay. Film samples were placed in a quartz cell with chloroform-heptane mixture at the bottom and then illuminated with 345 nm excitation light; P_y fluorescence emission was detected at 395 nm. Figure 1 shows the fluorescence decay profiles of P_y at various film formation steps. It is observed that as the vapor exposure time, t_e, increased, excited P_y decayed faster, thereby indicating that the quenching of excited P_y increases. Here the role of the solvent molecules is to add the quasi-continuum of states needed to satisfy the energy resonance conditions; i.e., the vapor molecules act as an energy sink for rapid vibrational relaxation, which occurs after the rate-limiting transition from the initial state.

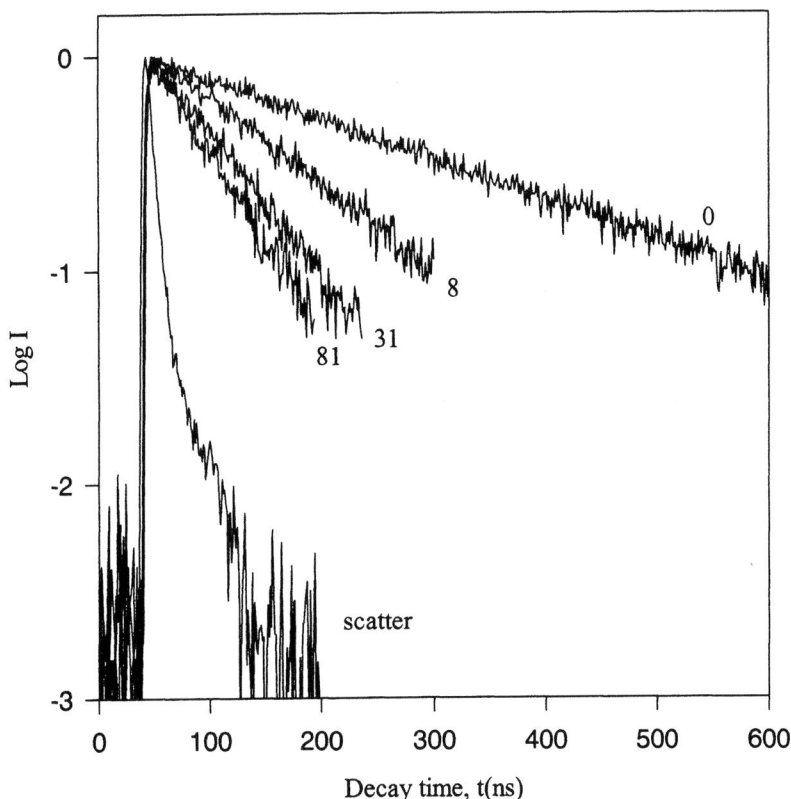

Figure 1. Fluorescence decay profiles of P_y at various film formation steps. The number on each curve represents the vapor exposure time, t_e, in minutes. The sharp peak represents the lamp of SMS.

To probe the vapor-induced film formation processes, the fluorescence decay curves were measured and fitted to Equation (1) (30). A and τ values were produced at each film formation step using linear least-squares analysis. In Figures 2a and b the measured τ values for the film formation experiments performed at 50 and 80 % chloroform content, are seen to decrease exponentially as t_e is increased. To quantify these results, a Stern-Volmer type of quenching mechanism is proposed for the fluorescence decay of P_y in latex film during vapor-induced film formation, where Equation (4) can be employed. For low quenching efficiency, $\tau_o k_q [M] < 1$, Equation (4) becomes

$$\tau \approx \tau_o \left(1 - \tau_o k_q [M]\right) \tag{5}$$

Here $[M]$ is the quencher (vapor) concentration at time t_e. The relation between the lifetime of P_y and $[M]$ in film can be obtained approximately using the volume integration of Equation (5) and the relation

$$\frac{\langle \tau \rangle}{\langle \tau_o \rangle} = 1 - C \frac{M}{M_\infty} \tag{6}$$

where $C = \tau_o k_q M_\infty / \upsilon$. Here υ is the volume of the film. The vapor sorption is calculated by integration $[M]$ over the differential volume (30). M_∞ is the vapor sorption at infinity. Combining Equation (6) with Prager-Tirrell's crossing density equation (31) and assuming that vapor penetration is proportional to the chain interdiffusion due to plasticization, i.e., the increase in vapor sorption causes the increase in crossing density, the relation

$$1 - \frac{\langle \tau \rangle}{\langle \tau_o \rangle} = Bt^{1/2} \tag{7}$$

where B is the related constant, includes the reptation frequency, υ of the PMMA chains. The fit of the data in Figures 2a and b to Equation (7) produced B values, in which the only variable is the reptation frequency, υ. B values are plotted versus percentage of chloroform in Figure 3, where it is seen that as percentage chloroform is increased, B values increase. This behavior can be explained by proposing that under high percentage chloroform vapor exposure, polymer chains reptate at higher frequencies during film formation processes, which result in large B values. Most probably large B values are caused by high crossing densities due to fast reptation of polymer chains.

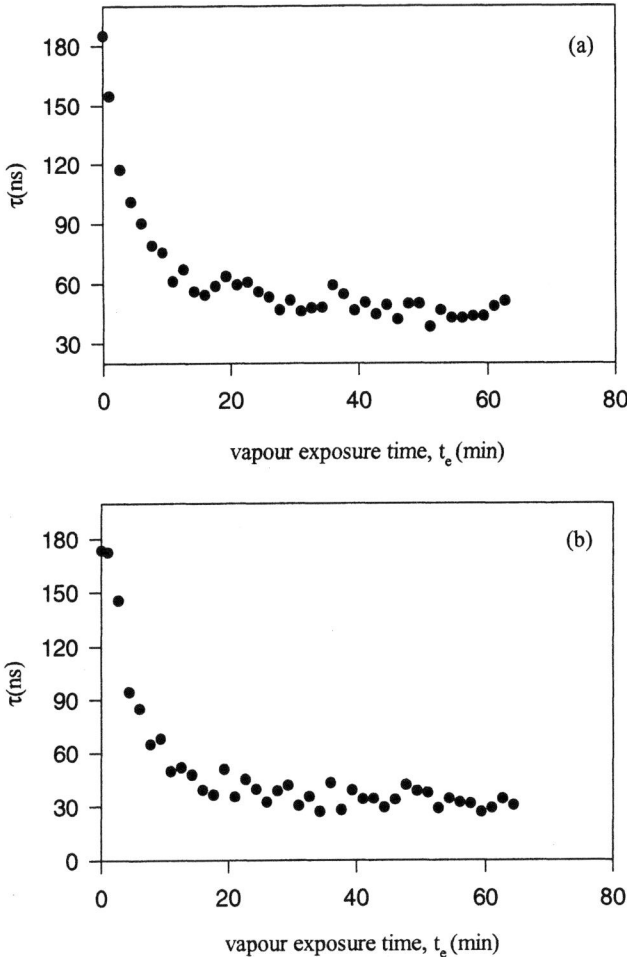

Figure 2. Plots of τ values versus vapor exposure time, t_e, for the film formation experiments performed with a- 50 % and b- 80 % chloroform content.

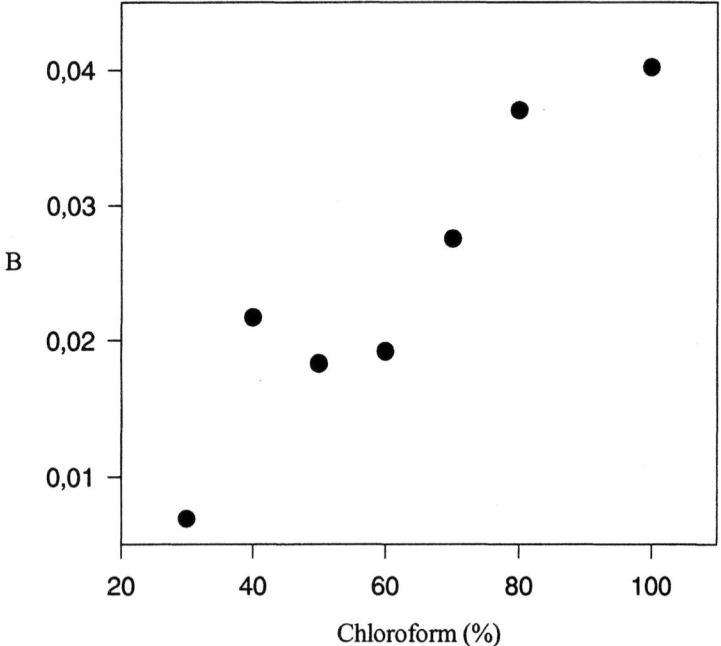

Figure 3. Plot of B values versus percentage chloroform in the solvent mixture.

For the DET measurements latex film preparations were carried out in the following manner; The same weights of naphthalene (N) and P_y particles were dispersed in heptane in a test tube for which solid content was 0.24 %. Film samples were prepared from this dispersion by placing various numbers of drops on 3×0.8 cm^2 glass plates and allowing the heptane to evaporate. The average film thickness was estimated to be approximately 20 μm. A LS-50 Perkin Elmer Spectrophotometer was used for SSF measurements. For the vapor induced latex film formation experiments (32), various solvents were chosen with different solubility parameters. Spectroscopically pure grade ethylbenzene (EB), toluene (TO), ethylacetate (EA), benzene (BE), chloroform (CH), dichloromethane (DM), terahydrofuran (THF) and acetone (AC) were purchased from Merck and used as received. Film formation experiments were performed in a 1×1 cm^2 quartz cell under solvent vapor. This cell was placed in the spectrophotometer, and fluorescence emission spectra were monitored at a 90° angle. Film samples were attached to one side of the quartz cell that contains a small amount of solvent at the bottom; the film was then illuminated with 286 nm of excitation light. Peak heights of N- and P_y-emission spectra were monitored during the vapor-induced film formation process. Seven different film formation experiments were run for the given solvents. Figure 4 presents the fluorescence emission spectra of N-P_y films when they are excited at 286 nm during the film-formation process induced by vapor of benzene. It is observed that P_y intensity, I_{P_y}, increases as N intensity, I_N, decreases, indicating that energy transfer from N to P_y takes place during vapor induced film formation. In other words, P_y-and N-labeled PMMA chains interdiffuse during film formation, and as a result, the excited naphthalene (N*) molecule transfers its energy to the P_y molecule. In the mean time, scattered light intensity, I_{sc}, at 286 nm decreases, indicating the creation of more homogeneous and transparent film.

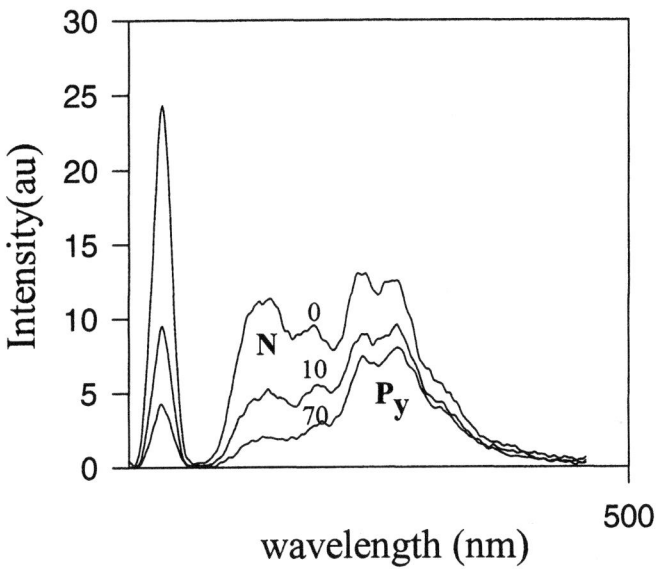

Figure 4. Fluorescence emission spectra of naphthalene (N) and pyrene (P_y) during organic vapor-induced latex film formation in benzene. Peaks at the left-hand side of the spectra represent the scattered light intensity during film formation processes. The number on each spectra denotes the vapor exposure time in minutes.

To present the evolution of energy transfer, that is, film formation induced by vapor, I_{P_y}/I_N ratios are plotted versus time of film formation. Figure 5 presents I_{P_y}/I_N versus time plots for benzene. Where it can be seen that all curves increase as the vapor exposure time is increased and reaches a plateau $(I_{P_y}/I_N)_{st}$ in an almost similar fashion at greater lengths of time. The chloroform sample reaches a plateau much faster than the others. Indicating that chain interdiffusion occurs much faster in this sample than in the other. Here it is interesting to note that all I_{P_y}/I_N curves start around 1, which may indicate that there is some energy transfer between surface dyes at 0 time. At this stage, an immediate conclusion can be reached by taking into account the solvent

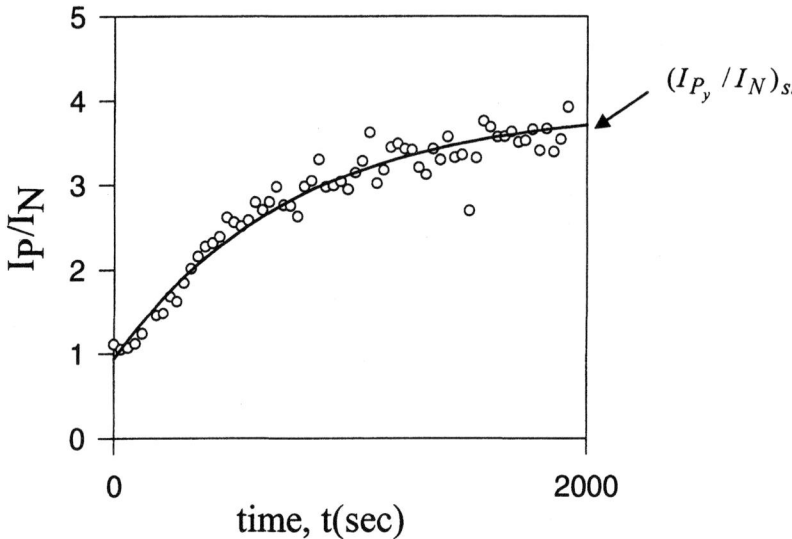

Figure 5. Plot of pyrene to naphthalene intensity ratio I_{P_y}/I_N versus vapor exposure time, t (film formation time) for benzene. $(I_{P_y}/I_N)_{st}$ indicates the saturation point of I_{P_y}/I_N.

characteristics such as molar volume, V. The plot of $(I_{P_y}/I_N)_{st}$ versus V is presented in Figure 6, where it can be seen that as the molar volume, V, increases, $(I_{P_y}/I_N)_{st}$ decreases. As $(I_{P_y}/I_N)_{st}$ is defined as the extent of energy transfer between N and P_y, small values of $(I_{P_y}/I_N)_{st}$ indicate to low energy-transfer values. In short, Figure 6 presents the extent of energy transfer between dye molecules at the equilibrium state of vapor uptake. Here it is observed that solvent molecule (EB) with largest V shows a lower energy transfer from N to P_y. However, small solvent molecules such as DC and AC present a higher amount of energy transfer from N to P_y at the equilibrium state of vapor-induced film formation. Here, most probably, large molecules create larger free volumes that prevent energy transfer from N to P_y because of longer distance between them. However, smaller free volumes created by small molecules cannot affect the energy-transfer process between N and P_y.

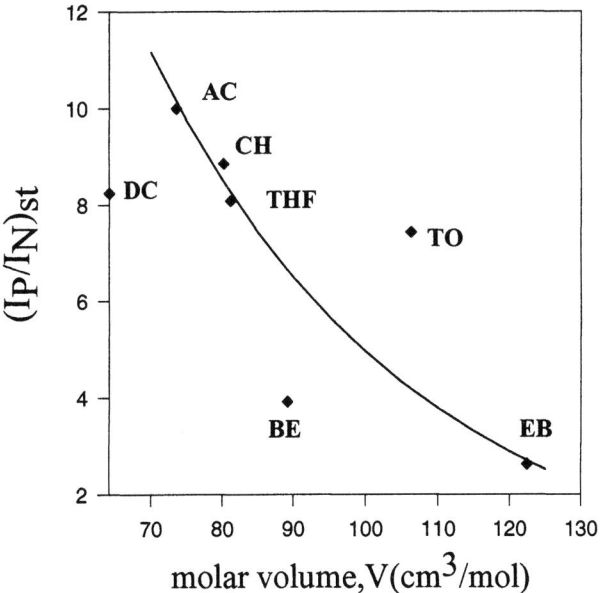

Figure 6. Plot of the plateau value, $(I_{P_y}/I_N)_{st}$ versus molar volume, V of the solvent molecules.

To quantify the results in Figure 5, the following mechanisms has to be considered, when the incident light intensity, I_o excites the N molecule in the latex film

$$N^* + P_y \xrightarrow{k_{ET}} N + P_y^* \qquad (8)$$

$$N^* \longrightarrow N + h\nu_N \qquad (9)$$

$$P_y^* \longrightarrow P_y + h\nu_P \qquad (10)$$

where N^*, P_y^* are the excited naphthalene and pyrene molecules which emit photons and k_{ET} is the DET rate constant. The time dependent rate equation that describe the above mechanism for N^* and P_y^* can be written as

$$\frac{d[N^*]}{dt} = I_o A_N - (k_{fN} + k_{AN})[N^*] - k_{ET}[P_y][N^*] \qquad (11)$$

$$\frac{d[P_y^*]}{dt} = I_o A_P - (k_{fP_y} + k_{AP_y})[P_y^*] + k_{ET}[P_y][N^*] \qquad (12)$$

where A_N and A_P are the Beer's Law absorbances and k's are shown in Equation (2). The solution of Equation (11) and Equation (12) for the steady state hypothesis produce the following equation (32)

$$\frac{I_{P_y}}{I_N} = C[P_y] \qquad (13)$$

where C is the related constant for N-P_y pairs. As N is excited at 286 nm, the above assumption for $A_P = 0$ is quite reasonable. Integrating Equation (13) over the total volume, V, of latex film and assuming that the total number of P_y-labeled chains, crossing the interface is proportional to the crossing density, then the following relation can be produced (32):

$$\frac{I_{P_y}}{I_N} = C'v^{1/2}t^{1/2} \qquad (14)$$

here, C' is a related constant and v is the back-and-forth frequency for reptating polymer chain in the latex film.

The normalized $(I_{P_y}/I_N)_{nr}$ form of the data in Figure 5 are plotted according to Equation (14). Fits are nice, and support the Prager-Tirrell model, where chain transport obeys the $t^{1/2}$ time dependence. The slope of the straight lines produces $C'v^{1/2}$ values, which are plotted in Figure 7 against solubility parameters, $(\delta - \delta_P)$. Here one immediately thinks that polymer-solvent interactions are responsible for the interdiffusion processes during vapor induced film formation. It is well known that solution theory predicts that the polymer-solvent interaction parameter is related to the solubility parameter, δ, and the molar volume, V, via the following relation (33):

$$\chi = \frac{V}{RT}(\delta - \delta_P)^2 \qquad (15)$$

where R is the gas constant, T is the temperature and δ_P is the solubility parameter of the polymer. This theory leads to the conclusion that polymer material swell and that chains interdiffuse in small molecular liquid only if $(\delta - \delta_P)$ is very small. The plot in Figure 7 shows that $(\delta - \delta_P)$ approaches zero when $\sqrt{C'v}$ value is very large for the CH molecule, where $\delta_P = 9.3(cal/cm^3)^{1/2}$ was taken for PMMA. From here one can conclude that the reptation frequency, v, is strongly correlated to the polymer-solvent interaction parameter, χ, which results in quick chain interdiffusion during vapor-induced latex film formation.

Photon Transmission Studies

Molecular Weight Effect on Latex Film Formation

Transparencies of the films formed from low molecular weighted (LM) and high molecular weighted (HM) latexes were studied by measuring the transmitted photon intensities, I_{tr} by using UV-Visible (UVV) spectrophotometer. Two

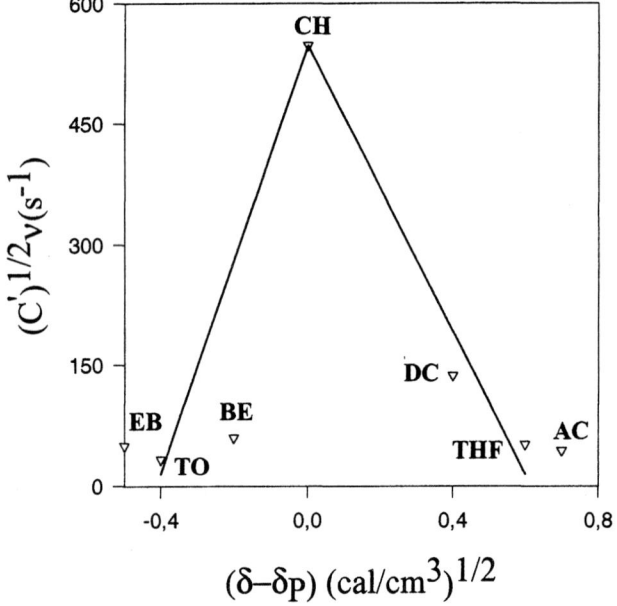

Figure 7. Plot of the relation between back-and-forth frequencies, υ, and solubility parameters, $(\delta - \delta_P)$.

different batches of PMMA particles were prepared separately in a two-step process. First MMA was polymerized to low conversion in cyclohexane in the presence of poly(isobutylene) (PIB) containing 2 % isoprene units to promote grafting. The graft copolymer so produced served as a dispersant in the second stage of polymerization, in which MMA was polymerized in a cyclohexane solution of the polymer. Details have been published elsewhere (*34*). A combination of ^{1}H-NMR and UV analysis indicated that these particles contain 6 mol % PIB and DSC results show that glass transition temperature, T_g of these particles (HM and LM) are both found to be around 390 °K. The particle size of LM and HM particles are measured using scanning electron micrographs and found to be 2 μm and 0.5 μm respectively. In the first and second batch of particles, molecular weights of graft PMMA were measured as $M_w=2.15\times10^5$ and $M_w=1.1\times10^5$ respectively. These particles are used to prepare HM and LM samples respectively. The polydispersities of the corresponding PMMA were 1.49 and 2.33 for the HM and LM particles. Two different sets of films were prepared from the dispersions of HM and LM particles in heptane by placing same number of drops on a glass plates with the size of 0.9×3.2 cm^2. Each set of samples contains seven different films. Annealing process of the latex films were performed in an oven in air above T_g of PMMA after evaporation of heptane, in 60-, 30-, 15-, 10-, 5-, 2.5- and 1-min time intervals at elevated temperatures between 383-483 °K and 383-543 °K for LM and HM film samples respectively. UVV experiments were carried out with the model Lambda 2S UV-Visible spectrometer from Perkin-Elmer and transmittance of the films was detected between 300-400 nm. All measurements were carried out at room temperature after the annealing processes were completed.

Transmitted photon intensities, I_{tr} from the LM and HM latex films has shown that films become transparent when they are annealed. In other words films scattered less light due to homogenization during film formation. When the I_{tr} intensities are compared for the LM and HM film samples, it is seen that the HM film needs higher annealing temperatures to reach the same transparency as that of the LM film for the same time intervals. The increase in I_{tr} may be interpreted by the mechanisms of void closure, healing and interdiffusion processes respectively (*35*). Spherical particles that have increasing surface energy flows to intervoids (void closure) at the early stage of annealing where the radius of interparticle voids becomes smaller and film surface becomes more homogeneous, consequently transparency of film starts to increase. If the annealing is carried on, chain segments (minor chains) move across the particle-particle interfaces and therefore latex film becomes more transparent. This process is called healing. At high annealing temperatures the chains gain sufficient kinetic energy to transform its center of mass across the junction surface (interdiffusion) and therefore latex film becomes fully transparent. In order to have the better feeling for the possible mechanisms, the transmitted photon intensities, I_{tr} from the LM films are plotted versus annealing temperature for 1-, 5- and 60-min time intervals in Figure 8. It is seen that all I_{tr} intensity curves start to increase at different temperatures depending on their annealing times. Figure 6 presents two distinct linear regions. These regions can be

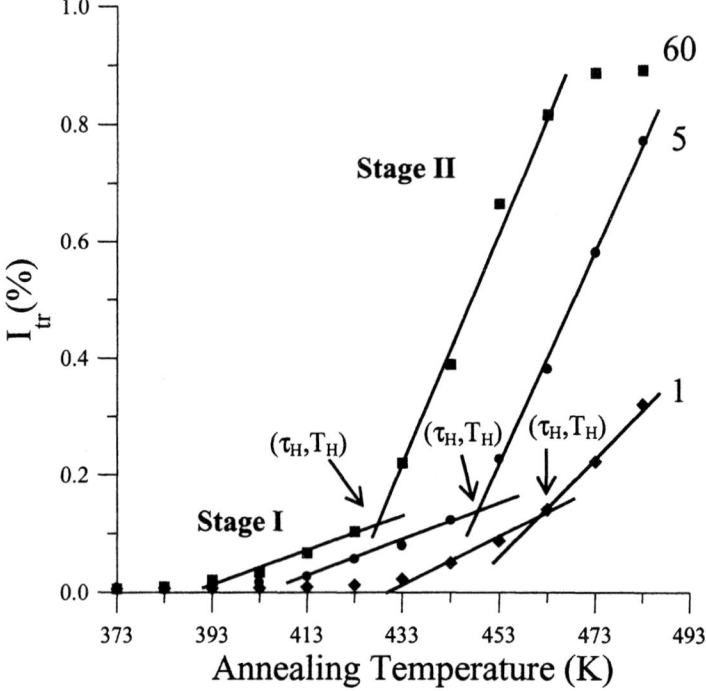

Figure 8. Plot of I_{tr} versus annealing temperature for 1, 5 and 60 min time intervals for LM latex films. Stages I and II present the void closure and interdiffusion processes respectively.

explained by the void closure and interdiffusion mechanisms during film formation process. Intersections between the two broken lines in Figure 8 presents the healing (τ_H, T_H) points (35).

In order to quantify the behavior of I_{tr} at the early stage of annealing, a phenomenological void closure model (36) can be introduced. Particle deformation and void closure between particles can be induced by shearing stress which is generated by surface tension of polymer i.e. polymer air interfacial tension. The void closure kinetics can determine the time for optical clarity and film formation. An expression to relate the shrinkage of spherical void of radius, r, to the viscosity of surrounding medium, η, was derived and given by the following relation (36).

$$\frac{dr}{dt} = -\frac{\gamma}{2\eta}\left(\frac{1}{\rho(r)}\right) \tag{16}$$

where γ is the surface energy, t is time and $\rho(r)$ is the relative density. It has to be noted that here surface energy causes a decrease in void size and the term $\rho(r)$ varies with the microstructural characteristics of the material, such as the number of voids, the initial particle size and packing. Here $\rho(r)$ can be defined as a volume ratio of polymeric material to voids where as r goes to zero $\rho(r)$ increases, however for large r values $\rho(r)$ decreases. Equation (16) is quite similar to one which was used to explain the time dependence of the minimum film formation temperature during latex film formation (*37*). The dependence of the viscosity of polymer melt on temperature is affected by the overcoming of the forces of macromolecular interaction which enables the segments of polymer chain to jump over from one equilibration position to another. This process happens at temperatures at which free volume becomes large enough and is connected with the overcoming of the potential barrier. The height of this barrier can be characterized by free energy of activation, ΔG during viscous flow. It is known that $\Delta G = \Delta H - T\Delta S$, where ΔH is the activation energy of viscous flow i.e. the amount of heat which must be given to one mole of material for creating the act of a jump during viscous flow. ΔS is the entropy of activation of viscous flow. With these considerations and assuming that the interparticle voids are in equal size and number of voids stay constant during film formation (i.e. $\rho(r) \propto r^{-3}$), integration of Equation (16) gives the relation (*38*)

$$t = \frac{2AC}{\gamma} \exp(\frac{\Delta H}{kT})(\frac{1}{r^2} - \frac{1}{r_o^2}) \qquad (17)$$

Where, C is a constant related to relative density $\rho(r)$ and r_o is the initial void radius. It is well established that decrease in void size causes an increase in mean free path of a photon which then results an increase in I_{tr} intensity (*38*). Then the assumption can be made that I_{tr} is inversely proportional to the void radius, r and Equation (17) can be written as

$$I_{tr}(T) = S(t)\exp(-\frac{\Delta H}{2kT}) \qquad (18)$$

Here r_o^{-2} is omitted from the relation since it is very small compared to r^{-2} values after void closure processes start. Where $S(t) = (\gamma / 2AC)^{1/2}$. $\ln I_{tr}$ versus T^{-1} plots of the data in Figure 8 presented two distinct regions, where Stages I and II present void closure and interdiffusion processes respectively. Stage I are fitted to Equation (18) and ΔH values are obtained and are plotted versus annealing time intervals in Figure 9 for LM films. The averaged ΔH values were found to be 150 kJ/mol and 134 kJ/mol for LM and HM films respectively.

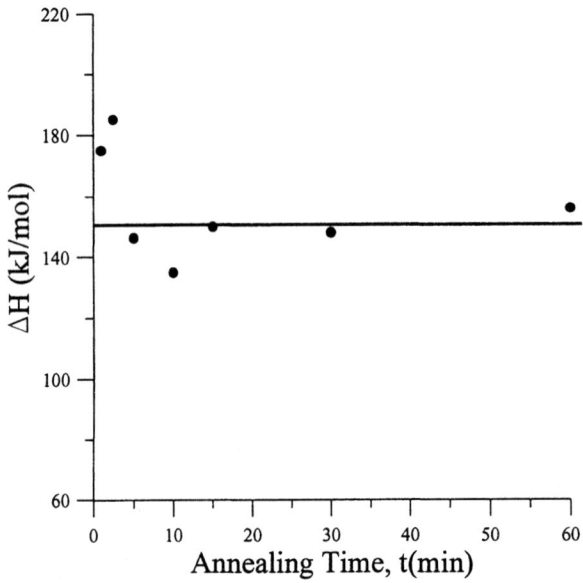

Figure 9. Plot of viscous flow activation energies (ΔH) versus annealing time (t) for LM films. ΔH values were obtained by fitting the stage I data to Equation (18).

When film samples were annealed at elevated temperatures for various time intervals above the (τ_H, T_H), a continuous increase in I_{tr} intensities was observed until they become saturated. This further increase in I_{tr} (stage II) can be explained by the increase in transparency of latex film due to the disappearance of particle-particle boundaries known as interdiffusion. As the annealing temperature is increased, some of the polymer chains cross the junction surface and particle boundaries disappear, and as a result, the transmitted photon intensity I_{tr} increases. The increase in annealing temperature causes total transfer of polymer chains across the boundary, which results completely transparent film. In order to quantify these results, the Prager-Tirrell (PT) model (30) for the chain crossing density can be employed. The total "crossing density" $\sigma(t)$ (chains per unit area) at junction surface was calculated from the contributions due to chains still retaining some portion of their initial tubes, $\sigma_1(t)$ plus a remainder, $\sigma_2(t)$. Here the $\sigma_2(t)$ contribution (backbone motion) comes from chains which have relaxed at least once. In terms of reduced time $\tau = 2\nu t / N^2$ the total crossing density can be written as

$$\sigma(\tau)/\sigma(\infty) = 2\pi^{-1/2} \tau^{1/2} \tag{19}$$

Here N represents the number of freely jointed segments in a chain. In order to compare our results with the crossing density of the PT model, the temperature dependence of $\sigma(\tau) / \sigma(\infty)$ can be modeled by taking into account the following Arrhenius relation for the back and forth frequency

$$\nu = \nu_o \exp(-\Delta E_b / kT) \tag{20}$$

Here ΔE_b is defined as the activation energy for the backbone of polymer chain. In order to explain the behavior of I_{tr} at stage II in Figure 8 it is assumed that I_{tr} is proportional to the crossing density $\sigma(T)$ at the interface, then the phenomenological equation can be written as

$$\frac{I_{tr}(T)}{I_{tr}(\infty)} = R \exp(-\Delta E_b / 2kT) \tag{21}$$

where $R = (8\nu_o t / \pi N^2)^{1/2}$ is a temperature independent coefficient. The activation energies (ΔE_b) of backbone motion were produced by fitting the stage II data to logarithmic form of Equation (21). The observed ΔE_b values are plotted against annealing time in Figure 10 for LM films. Where it is seen that smaller

ΔE_b values correspond to the short annealing time intervals for LM films. This behavior can be explained with the following sentence. Short chains cross the interface at short annealing times, which need smaller energy to execute backbone motion. However long chains need longer annealing time intervals to cross the interface, which poses large activation energies (ΔE_b).

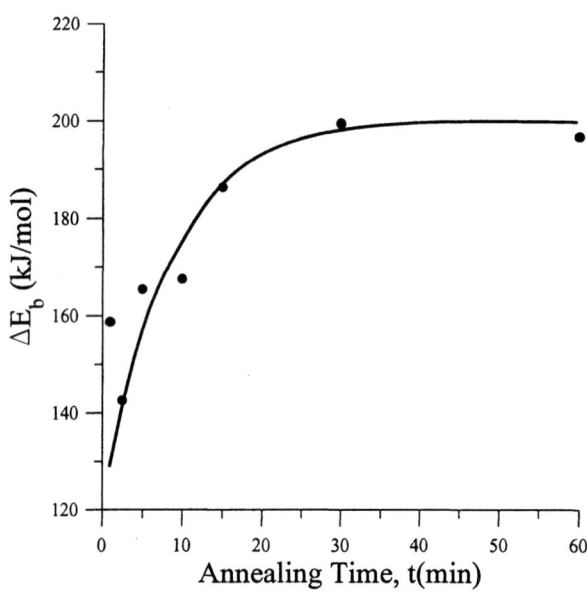

Figure 10. Plot of backbone activation energies (ΔE_b) versus annealing time (t) for LM films. ΔE_b values are obtained by fitting the stage II data to Equation (21).

Evolution in the transparency of the films prepared from HM and LM particles and exposed to the vapor of chloroform-heptane mixtures in various chloroform contents were monitored to study homogeneous film formation. When transmitted light intensities, I_{tr}, were monitored versus film formation time, t (i.e., vapor exposure time), for HM and LM latex films, it is seen that all I_{tr} intensity curves increase as t is increased (39). This behavior of I_{tr} suggests that the latex films become transparent to light as they are exposed to the solvent vapor. Small I_{tr} intensities were observed in low chloroform content experiments by indicating that vapor-induced film formation has not yet been completed in these samples. The I_{tr} increases faster as the chloroform content is increased. This behavior of I_{tr} intuitively predicts that powder film samples that have been exposed to high chloroform contents vapor create faster and more transparent films. In Figure 11 the film formation curves are compared, it is seen that HM films evolved much more slowly than LM films, most probably due to slow interdiffusion of high-molecular-weight polymer chains across the interface compared to low-molecular-weight chains. Here it has to be mentioned that most I_{tr} curves for HM films present shoulders at early times, which may be caused by the viscous flow of plasticized PMMA before the interdiffusion of chains has started (*39*). However LM films do not possess these shoulders. Most probably viscous flow occurs very fast for the LM system, before chain diffusion takes place. It is also observed that HM films need 100 % chloroform to complete film formation; however 40-60 % chloroform content is sufficient to do the same procedure in LM films.

The increase in I_{tr} can be attributed to the disappearance of particle-particle interfaces; i.e., as the exposure time of the solvent vapor is increased, more chains relax across the junction surface and, as a result, the crossing density increases. Now, one can assume that I_{tr} is proportional to the crossing density $\sigma(t)$ in Equation (19); then the phenomenological equation can be written as

$$\frac{I_{tr}(t)}{I_{tr}(\infty)} = A\, t^{1/2} \qquad (22)$$

where $A = (8\nu/\pi N^2)^{1/2}$. Here t is the film formation time, i.e., exposure time of the solvent vapor. The plots of Equation (22) for the I_{tr} data are shown in Figures 11a and b for HM and LM samples, respectively. The slopes of the linear relations in Figure 11 produce back-and-forth frequencies, ν, which are plotted against chloroform (%) for HM and LM films in Figure 12. Where it is openly present that polymer chains reptate faster as the chloroform percent is increased in both HM and LM film samples. It is also seen in Figure 12 that LM chains reptate faster than HM chains for a given chloroform content during vapor-induced film formation.

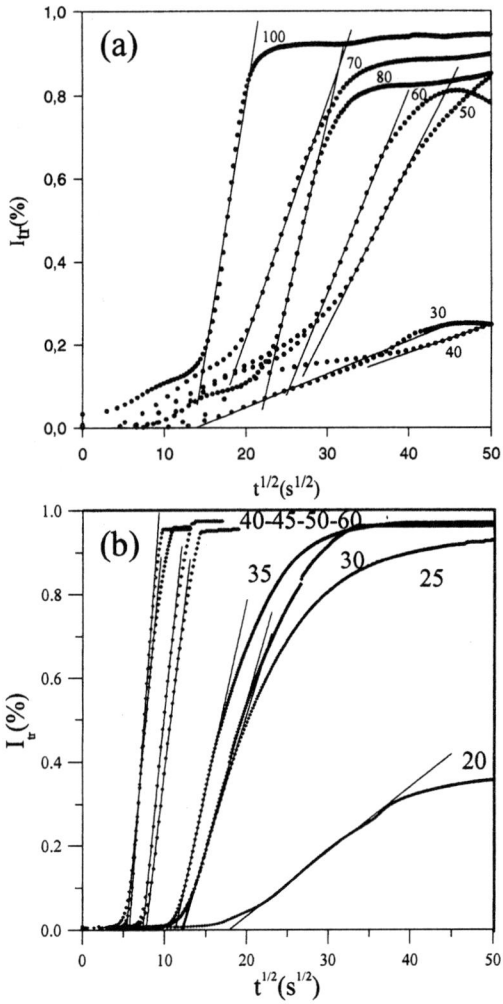

Figure 9. Plot of transmitted photon intensity, I_{tr} versus square root of film formation time, $t^{1/2}$, and fit of the data to Equation (22) for various chloroform contents for a- HM, b- LM films. Numbers on each curve indicate the chloroform volume (%) in the chloroform-heptane mixture.

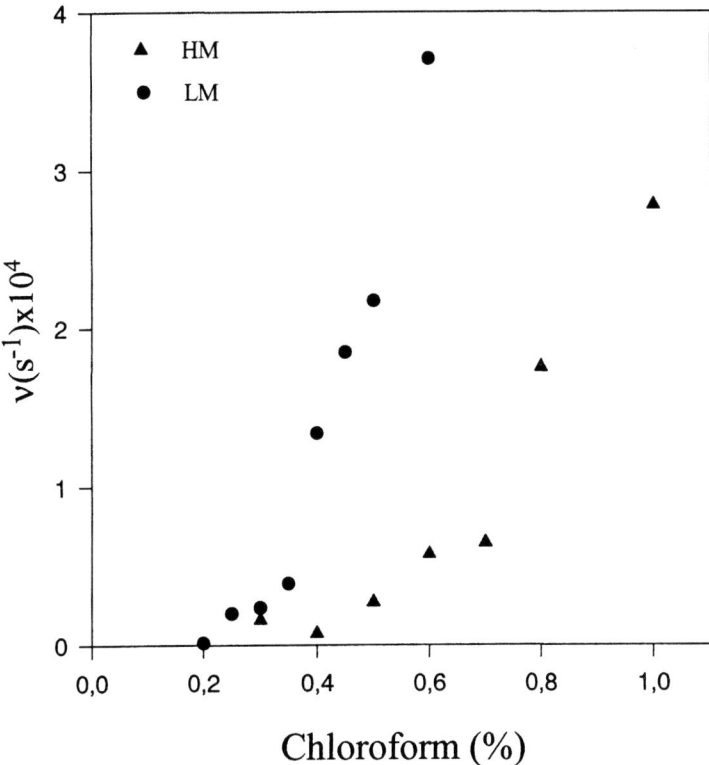

Figure 12. Comparison of back-and-forth frequencies, υ for HM and LM films versus chloroform (%).

Film Formation From Waterborne Nano-particles

Film formation from poly(methyl methacrylate-co-butyl methacrylate) (P(MMA-co-BMA)) nanosized latex particles which were produced by microemulsion polymerization was investigated using UVV technique (40). Latex film preparation was carried out in the following manner. P(MMA-co-BMA) latex particles were dispersed in water in a test tube. Six latex films were produced from this dispersion, by placing same number of drops on glass plates (with size of 0.9×3.2 cm^2) and allowing the water to evaporate at room temperature. Here, we were careful to ensure that the liquid dispersion from the droplets covers the whole surface area of the plates and remains there until the water has evaporated. Samples were weighed before and after film casting to determine the film thicknesses. The annealing processes of the latex films were performed in an oven in air above T_g of P(MMA-co-BMA) after the evaporation of water, in 5-, 10-, 15-, 20-, 30- and 45- min time intervals at elevated temperatures from 348 to 508 °K. The temperature was maintained within ±1 °C during annealing. After each annealing step transmitted photon intensity, I_{tr} from the film samples were detected between 400- 500 nm by UV-Visible (UVV) spectrophotometer (Lambda 2S of Perkin-Elmer, USA) at room temperature. A glass plate was used as a standard for all UVV experiments. The behavior of transmitted photon intensity, I_{tr} of latex films against annealing temperature for all time intervals, presented that films become transparent when they are annealed. In other words films scattered less light due to homogenization during film formation. The increase in I_{tr} may be interpreted by the mechanism of void closure, healing and interdiffusion processes respectively (40). Polymeric material in spherical particles that have increasing surface energy flow to intervoids (void closure) at the initial stage of annealing, as a result the radius of interparticle voids decrease and film surface becomes more homogeneous, consequently transparency of film starts to increase. If the annealing is carried on, chain segments (minor chains) move across the particle-particle interfaces (healing) and therefore latex film becomes more transparent. At high annealing temperatures the chains gain sufficient kinetic energy to transform its center of mass across the junction surface (interdiffusion) and therefore latex film becomes fully transparent.

$\ln I_{tr}$ versus T^{-1} plots of the data are presented in Figure 13 for 15 min annealing time interval. Figure 13 presents three distinct regions as predicted, where stages I, II and III present void closure, healing and interdiffusion processes respectively. Intersections between the broken lines show the void closure (τ_v, T_v) and healing (τ_H, T_H) points. It can be seen that (τ_v, T_v) and (τ_H, T_H) points all shift to the right hand side of the T^{-1} axis in Figure 13 i.e. left hand side of the T axis. This behavior indicates that for short annealing times films need higher temperature for void closure and healing processes to be occurred, otherwise it is reversed. Stage I in Figure 13 are fitted to Equation (18) and ΔH values are obtained from the slopes. Similarly produced ΔH values are listed in Table I for various annealing time intervals. The averaged ΔH value was found to be 66 kJ/mol. The intersections between stage I and II in Figure 13 correspond to the void closure point, (τ_v, T_v) which can be treated using Equation (17) as

$$\tau_v = S(r_v)\exp(\Delta H/kT_v) \qquad (23)$$

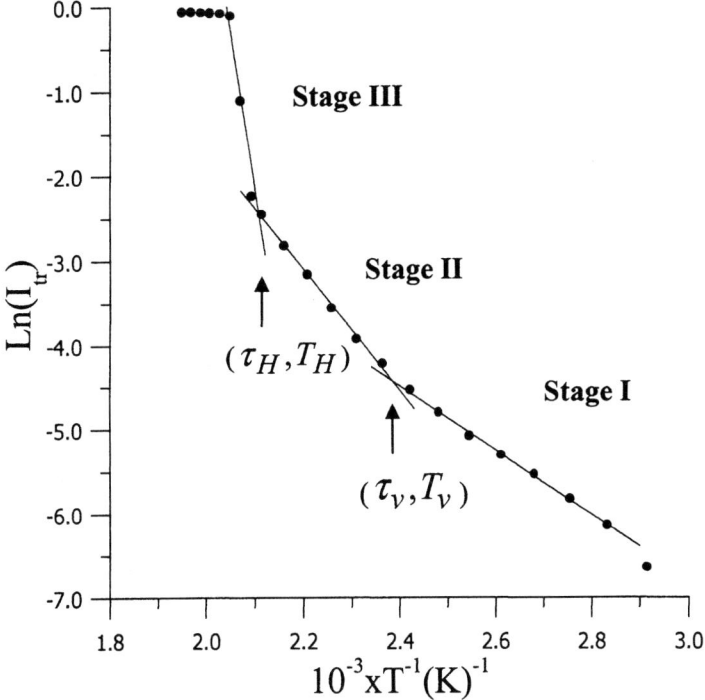

Figure 13. Logarithmic plots of the I_{tr} data versus the reverse of the annealing temperature (T^{-1}) for the films annealed at 15 min time intervals. Stages I, II and III represent the void closure, healing and interdiffusion stages respectively.

where $S(r_v) = 2AC/\gamma r_v^2$, here r_v is the minimal void radius at which optical path of a photon becomes longest in the latex film (*40*). As soon as r_v reaches zero, healing starts at the deformed particle-particle interfacesses. (τ_v, T_v) data are plotted in Figure 14 for the films annealed at 5, 10, 15, 20, 30 and 45 min time intervals, where it is seen that as the void closure time τ_v is decreased, the void closure temperature, T_v increases by obeying the Equation (23). The logarithmic form of Equation (23) is fitted to the data and ΔH is produced from the slope of the straight line and found to be 88 kJ/mol. The differences between ΔH values

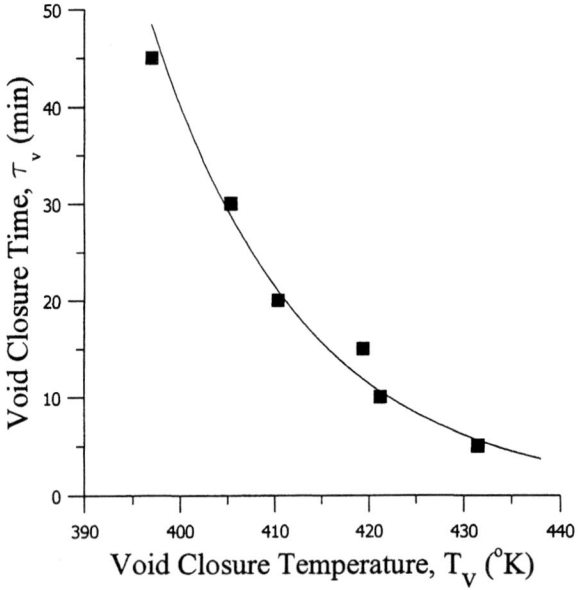

Figure 14. Plot of void closure points (τ_v, T_v), which were obtained from the intersection of the broken lines of stages I - II in Figure 13.

obtained from Equation (18) and Equation (23) most probably comes from the determination of the void closure points (τ_v, T_v).

In order to explain the behavior of I_{tr} in stages II and III, it is assumed that I_{tr} is proportional to the crossing densities $\sigma_1(T)$ and $\sigma_2(T)$ at these regions respectively, then phenomenological equation can be written

$$\frac{I_{tr}(T)}{I_{tr}(\infty)} = R\exp(-\Delta E/2kT) \qquad (24)$$

The activation energies of minor chain (ΔE_H) and backbone motion (ΔE_b) were produced by fitting the data in Figure 13 to logarithmic form of Equation (24) at the stages II and III respectively and are listed in Table I. Averaged value of ΔE_H and ΔE_b are found to be 114 kJ/mol and 552 kJ/mol respectively. Backbone activation

Table I. Experimentally measured activation energies of void closure ΔH, minor chains ΔE_H and backbone motion of chains ΔE_b for various annealing times.

	Annealing Time Interval (min)						
	5	10	15	20	30	45	Average
ΔH (kJ/mol)	67	67	63	59	67	71	66
ΔE_H (kJ/mol)	130	117	117	126	96	100	114
ΔE_b (kJ/mol)	691	532	628	561	448	452	552

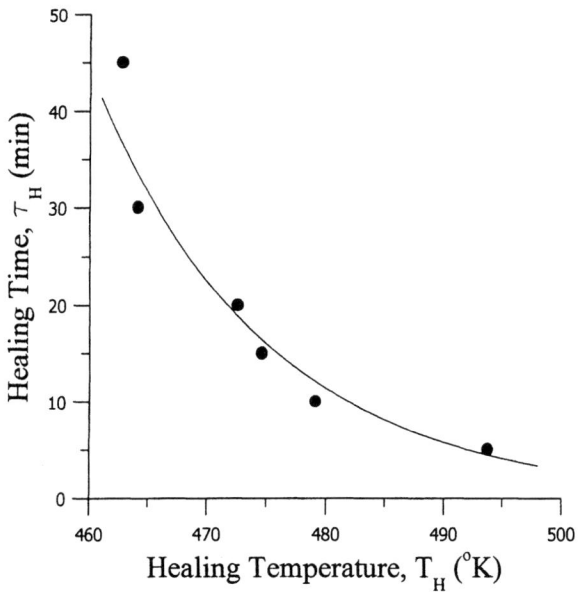

Figure 15. Plot of healing points (τ_H, T_H), which were obtained from the intersection of the broken lines of stages II - III in Figure 13.

energy (ΔE_b) was found to be five times larger than the activation energy (ΔE_H) for the minor chains. It is quite reasonable to accept that small chain segment (minor chain) needs much less energy to execute its motion than the energy needed to across the center of mass of a chain through the particle-particle interfaces. Intersections of stage II and III in Figure 13 are attributed to the healing point (τ_H, T_H), where the minor chain crosses the particle-particle interfaces. (τ_H, T_H) pairs are plotted in Figure 15, where it is seen that as τ_H is decreased T_H has to increase to execute the healing process using minor chains, during film formation.

Summary

In this chapter various latex film formation processes were studied using three different photophysical techniques. Time resolved and steady state fluorescence techniques were applied to elaborate vapor-induced latex film formation process, and then photon transmission technique was introduced to study molecular weight effect on latex film formation and film formation from waterborne nano-particles. Theoretical models were derived and used to understand void closure, healing and interdiffusion kinetics. Related parameters such as reptation frequencies, void closure, minor chain and backbone activation energies were determined.

References

1. Feng, J.; Winnik, M.A.; Shivers, R.R.; Clubb, B. *Macromolecules* **1995**, *28*, 7671.
2. Winnik, M.A.; Feng, J. *J. Coatings Technol.* **1996**, *68* (852), 39.
3. Pekcan, Ö. *Trends in Polym. Sci.* **1997**, *5* (6), 177.
4. Keddie, J.L. *Material Sci. and Eng.* **1997**, *21*, 101.
5. Niu, B.J.; Urban, M.W. *J. Appl. Polym. Sci.* **1998**, *70*, 1321.
6. Pekcan, Ö.; Arda, E. Interfacial behavior of latex films. *Encyclopedia of Surface and Colloid Science*; Marcel Dekker Inc., **2002**, 2691-2706.
7. Kim, K.D.; Sperling, L.H.; Klein, A. *Macromolecules* **1993**, *26*, 4624.
8. Mohammadi, N.; Klein, A.; Sperling, L.H. *Macromolecules* **1993**, *26*, 1019.
9. Whitlow, S.J.; Wool, R.P. *Macromolecules* **1991**, *24*, 5926.
10. Pekcan, Ö.; Winnik, M.A.; Croucher, M.D. *Macromolecules* **1990**, *23*, 2673.
11. Wang, Y.; Winnik, M.A. *Macromolecules* **1993**, *26*, 3147.
12. Wang, Y.; Zhao, C.L.; Winnik, M.A. *J. Chem. Phys.* **1991**, *95*, 2143.
13. Farinha, J.P.S.; Martinho, J.M.G.; Yekta, A.; Winnik, M.A. *Macromolecules* **1995**, *28*, 6084.
14. O'Neil, G.A.; Torkelson, J.M. *Macromolecules* **1997**, *90*, 5580.
15. Farinha, J.P.S.; Mantinho, J.M.G.; Kawaguchi, S.; Yekta, A.; Winnik, M.A. *J. Phys. Chem.* **1996**, *100*, 12552.

16. Feng, J.; Winnik, M.A. *Macromolecules* **1997**, *30*, 4324.
17. Winnik, M.A.; Wang, Y.; Haley, F. *J. Coating Technol.* **1992**, *64* (811), 51.
18. Juhue, A.; Lang, J. *Macromolecules* **1994**, *27*, 695.
19. Kim, K.D.; Sperling, L.H.; Klein, A.; Hammouda, B. *Macromolecules* **1994**, *27*, 6841.
20. Eckersley, S.T.; Rudin, A. *Prog. Org. Coat.* **1994**, *23*, 387.
21. Cannon, L.A.; Pethrick, R.A. *Macromolecules* **1999**, *32*, 7617.
22. Canpolat, M.; Pekcan, Ö. *Polymer* **1995**, *36*, 2025.
23. Keddie, J.L.; Meredith, P.; Jones, R.A.L.; Donald, A.M. *Macromolecules* **1995**, *28*, 2673.
24. Pekcan, Ö.; Arda, E. *J. Coating Technol.* **2001**, *73* (923), 51.
25. Arda, E.; Pekcan, Ö. *J. Colloid Interf. Sci.* **2001**, *234*, 72.
26. Steward, P.A.; Hearn, J.; Wilkinson, M.C. *Polym. Int.* **1995**, *38*, 1.
27. Chevalier, Y.; Pichot, C.; Graillat, C.; Joanicot, M.; Wong, K.; Maquet, J.; Lindner, P.; Cabane, B. *Colloid Polym. Sci.* **1992**, *270*, 806.
28. Birks, J.B. *Photophysics of Aromatic Molecules;* Wiley: New York, 1971.
29. Canpolat, M.; Pekcan, Ö. *Polymer* **1995**, *36*, 4433.
30. Akkök, B.; Uğur, Ş.; Pekcan, Ö. *J. Colloid Interf. Sci.* **2002**, *246*, 348.
31. Prager, S.; Tirrell, M. *J. Chem. Phys.* **1981**, *75* (10), 5194.
32. Pekcan, Ö.; Adıyaman, N.; Uğur, Ş. *J. Appl. Polym. Sci.* **2002**, *84*, 632.
33. Flory, P.J. *Principles of Polymer Chemistry;* Cornell Univ. Press: New York, 1953.
34. Pekcan, Ö.; Winnik, M.A.; Egan, L.; Croucher, M.D. *Macromolecules* **1983**, *16*, 699.
35. Arda, E.; Pekcan, Ö. *Polymer* **2001**, *42*, 7419.
36. Keddie, J.L.; Meredith, P.; Jones, R.A.L.; Donald, A.M. *Film Formation in Waterborne Coatings;* Provder, T.; Winnik, M.A.; Urban, M.W., Eds.; ACS Symp. Series; American Chemical Society: Washington, DC, 1996; *648*, p 332.
37. Mc Kenna, G.B.; Booth, C.; Price, C. *Comprehensive Polymer Science;* Pergamon Press: Oxford, 1989.
38. Pekcan, Ö.; Arda, E. *Colloid Surface A* **1999**, *153*, 537.
39. Akkök, B.; Pekcan, Ö.; Arda, E. *J. Colloid Interf. Sci.* **2002**, *245*, 397.
40. Arda, E.; Özer, F.; Pişkin, E.; Pekcan, Ö. *J. Colloid Interf. Sci.* **2001**, *233*, 271.

Chapter 10

Finite Element Modeling for Scratch Damage of Polymers

G. T. Lim, J. N. Reddy, and H.-J. Sue*

Department of Mechanical Engineering, Texas A&M University, College Station, TX 77843-3123
*Corresponding author: telephone: 979-845-5024; emsil: hjsue@tamu.edu

This chapter is concerned with the numerical simulation of scratch deformation on a polymer substrate by a semi-spherical indenter. A commercial finite element (FE) package, ABAQUS®, is used to study elasto-plastic scratch behavior on polymer surfaces. The study provides insightful mechanistic information that can be correlated to surface deformation and damage using material parameters. Of notable significance is that the FE analysis approach can be used to conduct parametric studies on material and geometric parameters that aid the construction of a physics-based model for describing the scratch behavior of polymers.

Introduction

Scratching phenomenon on a polymeric surface can be a complicated mechanical deformation process resulting in a material response that is still in need of further study. On the material response of polymers, scratches are often associated with permanently dented or fractured surfaces, material removal, fibrils drawing, and/or stress-whitening. Depending on the ductility and the operating environment [1], a polymer can undergo scratch induced damages, such as plastic yielding, crazing, microcracking, etc. To make the scratch analysis of polymers even more intractable, polymers are essentially viscoelastic. They will respond to deformation differently with time, temperature, stress state and strain rates and will strain-harden and strain-soften, depending on the material and the extent of deformation.

Scratch analysis of polymers can be made far more challenging by considering surface morphology variations caused by processing and the incorporation of fillers/additives in bulk materials for improved scratch performance. Injection molding, a key manufacturing process, is known to induce a skin-core morphology in the molded part [2, 3]. Due to the dissimilarity in the morphology, the skin layer will be expected to have different mechanical (scratch) response as compared to the core. The presence of filler particles, like talc, and additives, like lubricating agents in polymeric matrix can change scratch damage modes, which may in turn affect scratch visibility.

To gain better understanding on the scratch issue, researchers have relied mainly on the accumulated knowledge gained from the indentation study and propose new research methodologies to systematically delve into the scratch topic. Particularly on its mechanics aspect, the scope of study essentially covers viscoelasticity, plasticity or viscoplasticity, thermoelasticity, contact mechanics, tribology and dynamics. With the large deformation and the various material concerns, the analysis will involve both geometry and material nonlinearities. In view of the wide scope of scratch analysis and its nonlinearity, one can appreciate the level of difficulty involved to provide adequate analytical solutions for the problem. Even with reasonable simplifications, analytical solutions can only be limited to very simple cases.

To yield any forms of solution for the scratch problem, computational techniques established on sound engineering principles like the finite element method (FEM) [4] can be applied. However, the research effort on this approach remains scanty and is primarily focused on the study of indentation (see the well-compiled bibliography by Mackerle [5]). Tian and Saka [6] investigated the elastoplastic and plane-strain behavior of a layered substrate of bilinear materials under normal and tangential contact stresses using a commercial finite element (FE) ABAQUS® package. Their analysis however did not account for the dynamic effect of the moving indenter, and the contact between the indenter and substrate was not modeled. Another work that utilizes ABAQUS® for analysis is by Lee et al. [7], who modeled a steel ball scratching a rotating polycarbonate disk. While the material law adopted for the polycarbonate substrate is more realistic, they over-simplified a three-dimensional (3-D) problem to a two-dimensional plane-strain problem, rendering their FE analysis (FEA) to be non-applicable to their original problem, as mentioned in [8]. Bucaille et al. [9] and Subhash and Zhang [10] executed 3-D simulations of a displacement-controlled scratch deformation by a smooth rigid conical indenter on elastic-perfectly-plastic and bilinear materials, respectively. Their 3-D FEA are however unsuitable to study the scratch response of polymers since the material rheology adopted cannot capture the strain hardening and softening nature of polymers.

Learning from these earlier efforts, the present work is aimed at performing a 3-D contact analysis that has the capability to analyze the plastic flow induced under a scratch deformation. This work is set out to realistically model the scratch problem based on the test method and conditions standardized recently [11, 12]. This study will examine the mechanical response of polypropylene when scratched by a semi-spherical indenter using a commercial FE package, ABAQUS® (Version 6.3). Various physical and computational concerns outlining the FE modeling philosophy will be discussed. It will be shown that a better understanding of the deformation process of a scratch can be gained, and the FEA results can provide a plausible mechanistic explanation of the surface damage phenomena observed experimentally.

Physical and Computational Considerations for FE Modeling

Physics of a Scratch Deformation

When an object (indenter) makes a contact and traverses across the surface of another material (substrate), thereby making a scratch, the entire scratch process necessitates the consideration of several physical and material factors. Based on contact mechanics for indentation [13], the shapes of the indenter (*e.g.*, spherical, conical and others) will result in vastly dissimilar stress fields and different surface damage modes [14]. The size of indenter will in turn determine the scale of damage, from nanometer to millimeter. Since this work focuses on the mechanical response of polypropylene scratched by a stainless steel ball indenter with a diameter of 1 *mm*, the deformation will essentially remain in the millimeter scale. Besides the indenter, the geometry and shape of the substrate and the relative material property of the indenter and substrate will directly influence the extent of scratch damage. Detailed discussion will be provided in the subsequent sections on how these factors are included in the FEA.

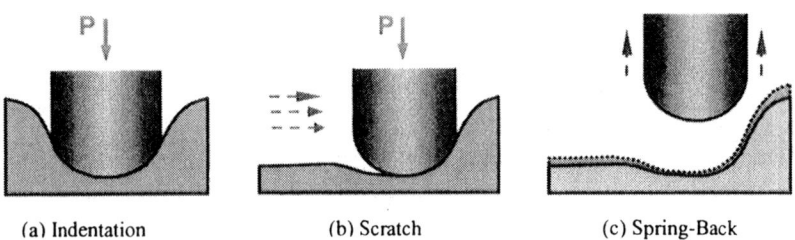

(a) Indentation   (b) Scratch   (c) Spring-Back

Figure 1. Various steps involved during a scratch process (load-controlled). (See page 4 of color insert.)

To model a scratch process, it is helpful to envision how a scratch occurs in an actual experimental setup. A scratch process can be separated into three mechanical steps (see Figure 1); the first is the indentation step whereby the indenter makes an indentation onto the substrate *via* a specific normal load or displacement. In the second scratch step, the indenter will plough through the surface and subsurface of the substrate and push or remove materials along the scratch path. For the final step, the indenter having come to a stop at the end of the scratch step is then lifted up from the scratch groove, thereby allowing some form of elastic recovery to take place in the substrate. This last step has commonly been known in the metal-forming industry as the *spring-back*.

Modeling Issues Related to Scratch

Dynamic Analysis of the Scratch Step

Due to the dynamical nature of the scratch step, a dynamic analysis is required. However, an immediate concern will be the analysis time and the resulting scratch speed of the indenter. For any realistic simulation, it is always ideal if the analysis can be performed over a time interval that mimics the actual physical process. Depending on the nature of the problem, the geometry, the type and size of element of the FE mesh, executing a realistic FE dynamic analysis can be time consuming and demands unrealistic computer resources.

In ABAQUS®/Explicit [15], an explicit scheme is employed to describe the time evolution of the independent variables. To ensure scheme stability and accuracy [4] of the dynamic analyses performed in this study, approximately one-tenth of the computed stable time increment was adopted. For an analysis time of 3 milli-second over a scratch length of 30 *mm* and running on a supercomputer (IBM® Regatta p690 with 1.3 GHz processors), an average computational time of four to five days and three gigabytes of memory were required. Though the resulting scratch speed may be fast for some applications, the study intends to examine the fundamental behavior of polymers under different loading and scratch conditions and the results, to be presented later, should still suffice to a better understanding of polymer scratch behavior.

Static Analyses of the Indentation & Spring-Back Steps

For both indentation and spring-back steps, they are not controlled in terms of time and will take on a longer time scale for completion. This is because during the indentation, the loaded indenter may have seated on the specimen for a while prior to scratching and the analysis of the surface damage does not follow immediately after the spring-back. Therefore, it is more appropriate to perform static analyses for these two steps. In this study, the indentation step is intended to be load-controlled, *i.e.*, the indentation is caused by a driving normal load acted on the indenter. However, due to the limitation of the contact algorithm of ABAQUS® [16], a firm contact has to be established first before a load can be specified correctly on the indenter. To do that, the indentation step is further divided into two steps [17]. In the first step, a displacement boundary condition is specified to push the indenter vertically onto the substrate and the normal reaction force will be noted; the indentation depth will be changed repetitively until it produces the desired normal reaction force. Once the desired force has been achieved, the displacement boundary condition is then replaced by a normal load of the same magnitude as the reaction force in the second step. For the spring-back step, it can be executed readily by prescribing a displacement boundary condition to move the indenter away from the surface of the substrate, thereby removing any contact with the substrate.

FE Mesh: Geometry, Element Type and its Boundary Conditions

To simulate the actual experimental conditions as discussed in [11, 12], the dimensions of the FE mesh follow closely to test specimens used and are taken to be 50 *mm* by 10 *mm* by 3 *mm*. However, compared to the test specimens, the length of FE mesh has been reduced to cut down computational time. Due to the symmetry of the model, there exists a plane of symmetry and this helps to reduce the size of the original mesh by half, as presented in Figure 2. For the indenter, the diameter of its spherical tip is taken to be 1 *mm*.

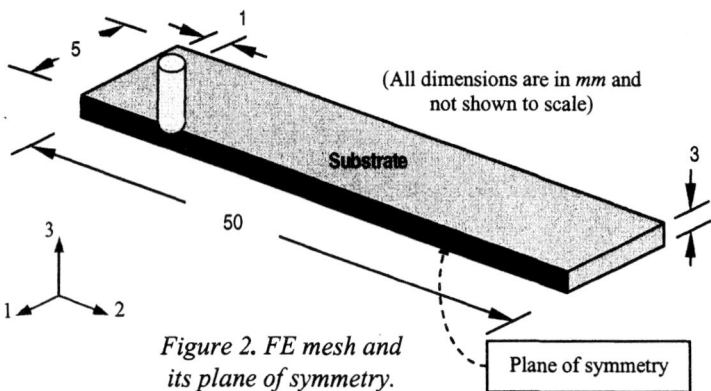

Figure 2. FE mesh and its plane of symmetry.

Since the aim of this study is to perform a 3-D analysis of the scratch process, 3-D linear (C3D8R) solid elements with hourglass mode control [15, 16] are utilized for their meshing simplicity. For accuracy, the FE mesh had been discretized to have 250 elements along its length, 40 elements along the width and 20 elements across the thickness. The indenter was modeled by a rigid analytical surface whose six degrees of motion is controlled by a reference node.

To mimic the clamping of test specimens at both ends [11, 12], all the nodes on both 1-3 planes of the FE mesh were restrained from movement in all three directions; the adopted coordinate system has been provided in Figure 2. Since test specimens are rested on the scratch machine, all the nodes on the bottom surface were restricted from moving in the vertical 3-direction. To impose the symmetry of the problem, the nodes along the plane of symmetry were not allowed to translate in the 1-direction. Since the original mesh has been reduced by half due to the symmetry, it should be noted that all loads imposed along the plane of symmetry should be scaled accordingly.

Material Law

The material considered for the substrate is polypropylene (PP). Like most polymers that are viscoelastic in nature, the constitutive behavior of PP will vary with strain rate and temperature. As a material input for FEA, the true compressive stress-strain curve of PP at a strain rate of 0.01/s has been adopted [18]. The density of PP is taken to be 905 kg/m^3, while its Young modulus is 1.65 GPa and Poisson ratio is 0.4. In the experimental setup of a scratch test [11, 12], the indenter tip is made up of stainless steel, which is more than a hundred times stiffer than PP and its yielding stress is about ten times as much. It is therefore equitable to treat and model the indenter as rigid and this assumption has been adopted for all analyses in this work.

Plastic Yielding Criterion

Since most polymers undergo strain softening and hardening at large plastic deformation, there will be a need to describe the onset and the evolution of plastic flow in the analysis. To predict and monitor any plastic deformation in our study, von Mises shear yielding criterion [19] had been employed. To vary the yield stress with the amount of plastic flow for describing material hardening and/or softening, the isotropic hardening rule was utilized [19].

Crazing, Debonding and Cracking Criterion

The fracture mechanisms that can lead to stress whitening are crazing, voiding, debonding, and cracking. Particularly for scratch damage, crazing should be treated as important as bulk shear yielding for several reasons. First of all, crazes are highly light-reflective in nature and if present, will increase scratch visibility on materials. Besides being a precursor of brittle cracking and fracture, crazing can occur at lower stress levels than that may require for bulk shear yielding [1]. Depending on materials, the state of deformation and the operating environment, crazing will compete against shear yielding to become the dominant fracture mechanism. It is therefore of research interest to look into the possible initiation of crazes during a scratch process. The criterion used for assessing craze initiation is also relevant to evaluate voiding, debonding, and cracking since they involve the same type of stress/strain components, *i.e.*, the critical strain and the maximum hydrostatic tension. Of the various criteria for craze initiation, the critical strain criterion by Bowden and Oxborough [20] is adopted for its sound physical basis and ability to account for a general triaxial state of stress. The criterion states that crazing occurs when the strain in any direction reaches a critical value and that this critical strain depends on the hydrostatic tension [20, 21]; mathematically, this criterion can be described as,

$$\varepsilon_C = A + \frac{B}{\sigma_1 + \sigma_2 + \sigma_3} \equiv \varepsilon_1 \qquad (1)$$

where ε_C is the critical craze strain and σ_i $(i = 1, 2, 3)^\dagger$ are the principal stresses while A and B are time-temperature-dependent parameters. In this study, ε_C is treated to be equivalent to the maximum principal strain, ε_1.

Contact Algorithm

To establish, track and maintain contact between surfaces, the contact pair (master-slave) algorithm for finite sliding as provided by ABAQUS® [15,16] was selected for the study. In this study, the analytical rigid surface of the indenter was assigned to be the master surface while the slave surface belonged to the top surface of the FE mesh for the substrate. Though the contact pair algorithm may be robust, the FE mesh must still be sufficiently refined to avoid any erroneous over-closure of contact surfaces. The FE meshes were created with the consideration that for the elements that come into contact with or near the indenter, the initial largest dimension of the smallest elements is not more than a quarter of the diameter of the indenter. Double precision calculation was used in the FEA to alleviate any contact noise that may compromise the results.

Surface Interaction

Like any two sliding surfaces, there bounds to be interaction between them in the form of friction and heat generation. As thermal effect will not be considered in this study, the heat generation between surfaces shall be ignored. For the frictional interaction between the surfaces, the basic Coulomb friction

---

† The average of the three principal stresses gives the hydrostatic stress.

model [22] had been incorporated in the FEA [15, 16] *via* the definition of the coefficient of adhesive friction. In all analyses performed, the coefficient of adhesive friction was taken to be 0.3.

Adaptive Remeshing

As the scratch process involves large deformation, the FEA may encounter convergence problem arising from severely distorted mesh. To maintain a high quality FE mesh throughout the analysis, adaptive remeshing, available in ABAQUS® [15], can be employed. Lagrangian adaptive meshing, which is suitable for transient problem with large deformation, had been selected to allow the adaptive mesh domain to move together with the material contained within. Adaptive remeshing is however highly computational intensive and hence time consuming. To reduce computational time, only mesh domains that are close to the scratch path had been assigned for remeshing.

Load Cases

Two load cases, *Load Cases A* and *B*, had been considered for this study, as summarized in Table 1.

Table 1: Load Cases A and B.

Load Case	Indentation Step		Scratch Step ($t = 3$ ms)	Spring-back Step
	Depth-controlled	Load Imposition		
A	$u_3 = -0.1258$ mm	$P = 15$ N*	$P = 15$ N $v = 10$ m/s	$u_3 = 3$ mm
B	$u_3 = -0.05401$ mm	$P = 5$ N	$P = 5 - 15$ N $v = 10$ m/s	$u_3 = 3$ mm

* Note that the normal load P has been scaled by half due to the plane of symmetry (see *FE Mesh: Geometry, Element Type and its Boundary Conditions*).

For Load Case A, the scratch process is characterized by a constant scratch speed v of the semi-spherical indenter under a constant normal load P. As for Load Case B, the normal load of the indenter increases linearly over the scratch length with a constant scratch speed. As mentioned earlier, the indentation process has been divided into two steps, namely the depth-controlled and load imposition steps. The analysis time of the scratch step for all analyses is set at 3 milli-second (*ms*) and the scratch length is 30 *mm*.

Results and Discussion

Scratch Depth and Width

Residual scratch depth and width are commonly adopted to quantify scratch damage experimentally [11, 12]. The residual scratch depth variation over the scratch length for the two load cases at the end of the spring-back step is plotted in Figure 3(a). As shown in Figure 3(a), a drastic increase in the scratch depth can be observed over the early portion of the scratch length. This increase can be attributed to the initiation of motion by the indenter that results in a sudden introduction of force inertia into the system. This added inertia will gradually cease to exist and is clearly illustrated by the scratch depth for Load Case A that settles into a steady state of approximately constant depth. Also, by comparing the scratch depths at the end of the scratch path where the normal loads and scratch speeds are the same for both load cases, the FEA results show that Load Case A will incur a more severe scratch damage, *i.e.*, a deeper scratch depth than Load Case B. This qualitative trend is also observed experimentally for scratch widths and depths by Wong et al. [12] where a comparison of experimental and FEA results was made, as shown in Figure 3(b).

Scratch Damage Process

Under controlled laboratory conditions, it is relatively easy to reproduce scratches on specimens. But without a precise and rapid video imaging capability, scratch tests in most experimental set-ups will occur too rapidly to capture the sequential mechanism for the formation of scratch grooves. For that, FEA is useful as it generates a database of results to capture the time evolution of the scratch process that can be reproduced graphically to aid visualization. Figure 4(a–d) shows the deformation sequence of the polypropylene substrate as the indenter ploughs through it. In Figure 4, layers of the substrate over a section of concern are shaded differently and the indenter is moving out towards the reader. Figure 4(a) first shows a relatively undeformed section of substrate that is ahead of the approaching indenter. The section begins to undergo compression and is squeezed upwards as the indenter moves ahead as shown in Figure 4(b). In Figure 4(c), the approaching indenter continues to exert its compressive action on the section while it also pushes the material sideways. Once the indenter overcomes and ploughs through the materials ahead, a scratch groove is formed, as illustrated by Figure 4(d).

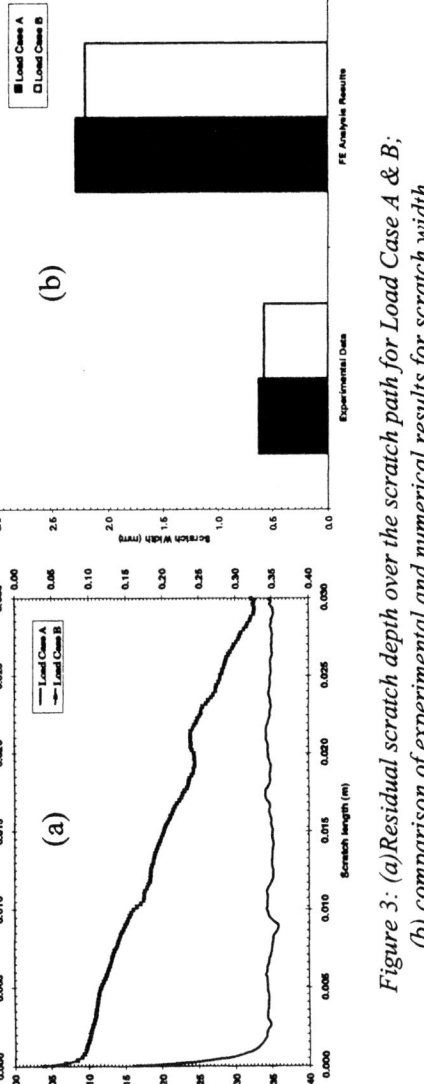

Figure 3: (a)Residual scratch depth over the scratch path for Load Case A & B; (b) comparison of experimental and numerical results for scratch width.

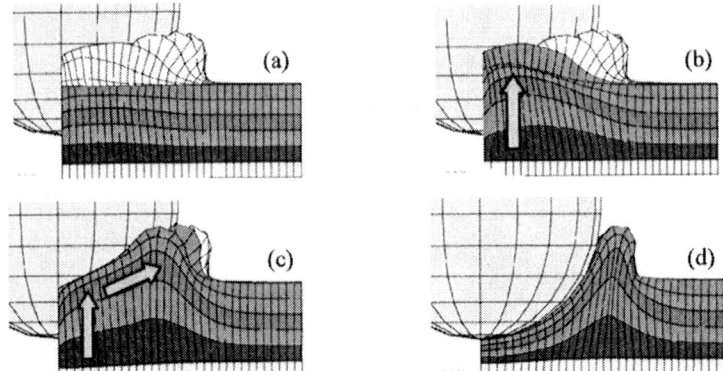

Figure 4: (a) Undeformed section; (b) section is compressed and squeezed upwards; (c) section is pushed to the side; (d) a scratch groove is formed. (See page 4 of color insert.)

Plastic Yielding & Crazing

As we know, shear yielding and crazing are two key damage modes for polymers. In the work of Lim et al. [23], numerical results have been reviewed for a phenomenological examination of the occurrence of damage in materials around the indenter tip and the formation of crazes during the scratch process. Henceforth, it is of research interest to look into the quantitative variation of plastic yielding and crazing along the scratch length. Presented in Figure 5(a) and (b) are the plots of the maximum envelope of the equivalent plastic strain and volumetric strain along the scratch path for Load Case B, respectively.

From Figure 5(a), it is evident to note the large extent of plastic damage occurring in the materials, especially towards the end of the scratch process. Though not being considered in the FEA, materials are likely to fracture and spall off from the scratch path beyond plastic strain levels of 200-300%. The consideration of ultimate material failure during scratching is currently pursued in future numerical works of the authors. For Figure 5(b), the use of volumetric strain allows one to correlate to the amount of crazing that takes place during the scratch process. Comparing both plots and considering the linear load increase in Load Case B, it is interesting to note that other than the initial portion, the variation of equivalent plastic strain follows a linear trend while the volumetric strain increases in a quadratic manner. This indicates that crazing will not be a dominant damage mode at the beginning of the scratch process but competes with shear yielding for dominance towards the end. Since crazing is characterized by the stress-whitening phenomenon, there will be a critical point for the volumetric strain to reach its threshold of stress-whitening. For future implementation, it is beneficial for the study of crazing damage if a criterion and a growth and failure mechanism for crazing can be incorporated in the FEA. Nevertheless, this study demonstrates the usefulness and its potential of the FEA to predict the onset of stress-whitening.

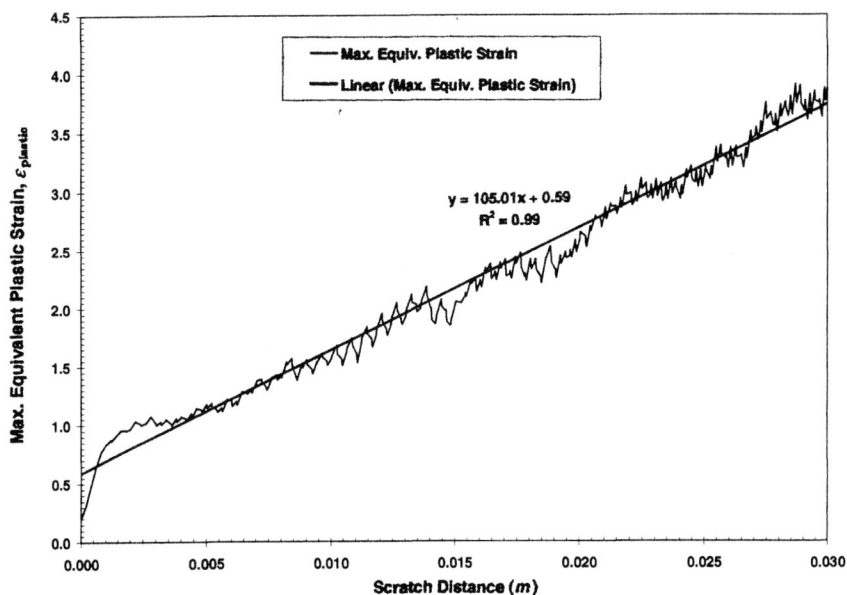

Figure 5(a): Maximum envelope of equivalent plastic strain along the scratch path.

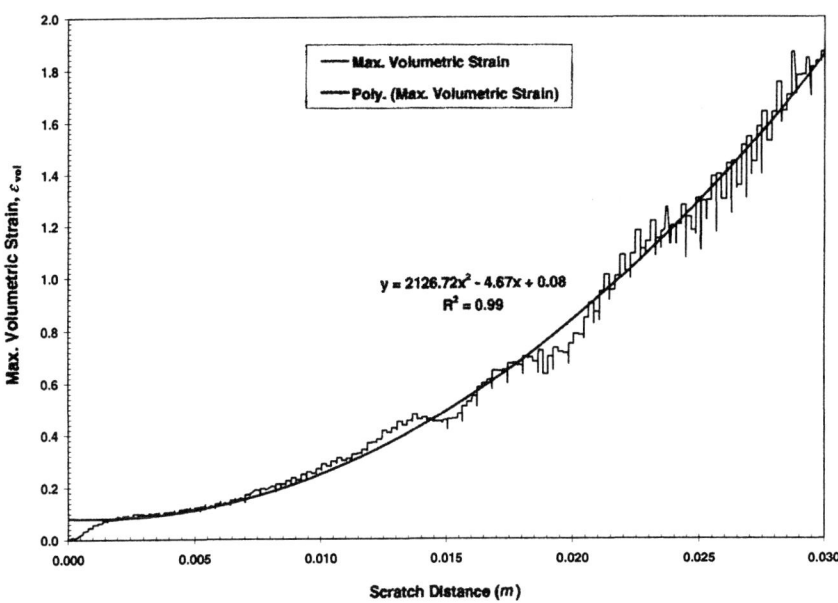

Figure 5(b): Maximum envelope of volumetric strain along the scratch path.

In spite of a relatively simple elasto-plastic material model and other assumptions adopted in this work, a good qualitative agreement between the FEA results and experimental observations is still being achieved, thereby establishing the usefulness of FEA in examining polymer scratch damage mechanisms for shear yielding and crazing/microcracking. Though there is yet a quantitative disparity between the FEA and experimental results, it can be overcome by implementing a more realistic material model and improving the FEA for a closer simulation of the actual experimental conditions.

Conclusions

In this chapter, a three-dimensional finite element analysis (FEA) was performed using a commercial package, ABAQUS® to study the mechanical response of a polypropylene substrate when scratched by a semi-spherical indenter. A detailed discussion of physical and computational considerations for the FEA of the scratch problem was presented. In addition, discussion of the FEA results was presented to elucidate the global response of the material (*i.e.*, scratch depth and width) during a scratch process. By examining the maximum principal stress in the finite elements around the indenter tip, a phenomenological description of the damage mechanism experienced by the scratched material was proposed. The examination of the global and local response of the material to the scratch deformation should provide the much-needed mechanistic understanding on the scratch response of polymers.

Acknowledgments

The authors would like to thank the financial support provided by the Texas A&M Scratch Behavior Consortium (Advanced Composites, Atlas-MTS, BP Chemical, Clorox, Luzenac, Solvay Engineered Polymers and Visteon) in this research endeavor. The authors would also like to acknowledge the financial support from the State of Texas (ARP #32191-73130) and Defense Logistic Agency (SP0103-02-D-0003). The second author acknowledges the support of the research by the Oscar S. Wyatt Endowed Chair. Special thanks are also due to the Society of Plastics Engineers – South Texas Section, for their generous donation of equipment for this research.

References

1. Kinloch, A.J.; Young, R.J. *Fracture Behavior of Polymers*; Elsevier Science Pub. Co.: New York, 1983.

2. López, L.C.; Cieslinski, R.C.; Putzig, C. L.; Wesson, R.D. Morphology characterization of injection moulded syndiotactic polystyrene. *Polymer* **1995**, *36(12)*, 2331-2341.
3. Pontes, A.J.; Oliveiria, M.J.; Pouzada, A.S. Studies on the influence of holding pressure on the orientation and shrinkage of injection molded parts. *ANTEC* **2002**, *1*, 516-520.
4. Reddy, J.N. *An Introduction to the Finite Element Method*; 2nd Ed., McGraw-Hill: New York, 1993.
5. Mackerle, J. Finite element and boundary element simulations of indentation problems - A bibliography (1997-2000). *Finite Elem. Anal. Des.* **2001**, *37(10)*, 811-819.
6. Tian, H.; Saka, N. Finite-element analysis of an elastic-plastic 2-layer halfspace - sliding contact. *Wear* **1991**, *148(2)*, 261-285.
7. Lee, J.H.; Xu, G.H.; Liang, H. Experimental and numerical analysis of friction and wear behavior of polycarbonate. *Wear* **2001**, *251(2)*, 1541-1556.
8. Wong, M.; Lim, G.T.; Rood, P.R.; Moyse, A.; Reddy, J.N.; Sue, H.-J. Scratch damage phenomena of polyolefin materials," *TPOs in Automotive Conference* **2003**.
9. Bucaille, J.L.; Felder, E.; Hochstetter, G. Mechanical analysis of the scratch test on elastic and perfectly plastic materials with the three-dimensional finite element modeling. *Wear* **2001**, *249(5-6)*, 422-432.
10. Subhash, G.; Zhang, W. Investigation of the overall friction coefficient in single-pass scratch test. *Wear* **2002**, *252(1-2)*, 123-134.
11. Wong, M.; Moyse, A.; Lee, F.; Sue, H.-J. Study of surface damage in polypropylene under progressive loading. *J. Mater. Sci.* **2004**, *39*, 3293-3308.
12. Wong, M.; Lim, G.T.; Moyse, A.; Reddy, J.N.; Sue, H.-J. A new test methodology for evaluating scratch resistance of polymers. *Wear* **2004**, *256(11-12)*, 1214-1227.
13. Johnson, K.L. *Contact Mechanics*; Cambridge University Press: UK, 1985.
14. Williams, J.A. Analytical models of scratch hardness. *Tribol. Intern.* **1996**, *29(8)*, 675-694.
15. ABAQUS®, Inc. *ABAQUS®/Explicit User's Manual Ver. 6.3* **2002**, *1-2*.
16. ABAQUS®, Inc. *ABAQUS®/Standard User's Manual Ver. 6.3* **2002**, *1-3*.
17. ABAQUS®, Inc. *ABAQUS®/Example Problems Manual Ver. 6.3* **2002**.
18. Arruda, E.M.; Azhi, S.; Li, Y.; Ganesan, A. Rate dependent deformation and semi-crystalline polypropylene near room temperature. *J. Eng. Mater. – T. ASME* **1997**, *119(3)*, 216-222.
19. Khan, A.S.; Huang, S. *Continuum Theory of Plasticity*; John Wiley & Sons: New York, 1995.
20. Bowden, P.B.; Oxborough, R.J. A general critical strain criterion for crazing in amorphous glassy polymers. *Philos. Mag.* **1973**, *28*, 547-559.

21. Argon, A.S.; Salama, M.M. Growth of crazes in glassy polymers. *Philos. Mag.* **1977**, *36(5)*, 1217-1234.
22. Bowden, F.P.; Tabor, D. *The Friction and Lubrication of Solids;* Claredon Press: Oxford, 1954.
23. Lim, G.T.; Wong, M.; Reddy, J.N.; Sue, H.-J. An integrated approach towards the study of scratch damage of polymer. *Journal of Coating Technology Research.* **2005**, in press.

Chapter 11

An Artificial Neural Network Approach to Stimuli-Response Mechanisms in Complex Systems

Eduardo Nahmad-Achar[1] and Javier E. Vitela[2]

[1]Centro de Investigación en Polímeros (CIP), Grupo Comex, 55885 Tepexpan, Edo. de México, México
[2]Instituto de Ciencias Nucleares, Universidad Nacional Autónoma de México, 04510 México, D. F., México

Artificial intelligent methods in the form of neural networks are applied to a set of different families of architectural alkyd enamels as a rational approach to the study of complex stimuli-response mechanisms: a feedforward neural network with sigmoidal activation functions was used with a conjugate gradient algorithm to recognize the complex input-output relation between the paint properties and the formula components. It is shown that good predictive power (of over 90%) is obtained within given appropriate uncertainty tolerances, and that the set of variables which most significantly affect a given property can also be determined through this methodology via a sensitivity analysis.

Introduction and Motivation

The reformulation of old coatings and/or development of new products in the search for better performance, more durability, lower costs, or merely to meet environmental constraints, is not only costly but very time consuming, due to the complexity of the system. If an item is changed in a coating formulation, it is important to ask how do other components and properties react; the answer, however, is far from trivial: we would need millions of experiments to describe the micro-dynamics of the system, as *ab initio* calculations are still very far fetched.

If y represents a measurable property of a system, and we can either write $y = f(x_1, ..., x_n)$ with $\{x_i\}_{i=1, ..., n}$ the independent variables of the system and f a known function, or there exists a computational algorithm through which y can be calculated, then one can study its behavior and even predict it. But in most situations in the real life f is unknown, and at best we have some independent sensors on the system to estimate y, whose inter-relations are complex and unknown. Such is the case of a paint formulation. An interesting example was given in (*1*) for a latex-based paint, whose measurable characteristics such as leveling, scrub resistance, adherence, gloss, etc. depend on the formula components both at a macroscopic and at a molecular level. Leveling, to take one of them, is supposed to be a function of minimum-film-formation temperature, particle size distribution, relative humidity, temperature, surface tension, the diffusion coefficient for each co-solvent present, film thickness, pigment volume concentration, critical pigment volume concentration, and various other system variables; but this function is unknown and so are its relationships with scrub resistance and adherence.

Although the encountered complexities preclude the development of mathematical models based on first principles, these systems can be studied via mechanisms which associate the output measurements of the sensors with known families or groups, i.e. which can construct associations based upon the results of known situations. If they can further adjust themselves (i.e. "learn") to newly presented situations, then we would be in a position to eventually generalize them. One such mechanism is an artificial neural network, with the added bonus of being able to provide the user with the most important input variables that affect a given property under study (in escence, a stimuli-response mechanism).

Artificial neural networks (ANN's) are self-learning computational "intelligent" techniques which do not require vast programming skills, can be structured in many various ways, and which consist in storing vast amounts of information as patterns and using those patterns to solve specific problems. They are so called because they are inspired by what is believed to be the way in which our brains learn from experience (although the exact workings of the brain are still unknown). Since, presently, the work of a paint formulator strongly depends on past experience, and since it has not been possible to predict the outcome through simple mathematical relations, it is only natural to try to apply ANN's to paint formulation. Well established for many years in academic environments, artificial intelligence methods in the form of computer-based technologies, presumptuously called "expert systems", have only very recently been applied to product formulation and/or process monitoring; mainly the latter (cf. e.g. *2 - 8*). Of these, few examples can be found in the coatings industry, mainly due to companies being reticent about revealing information. We believe, however, that it is precisely their complexity that makes them excellent candidates of study through self-learning mechanisms.

In this work we present the use of artificial intelligence methods in the form of supervised, feed-forward artificial neural networks as a tool for coatings (re)formulation. By applying them to a set of different families of architectural alkyd enamels, we show that good predictive power and discrimination is obtained for 2 studied properties, viz. specular gloss at 60° and drying time to touch. These were chosen not only because they are determinant for user needs, but also because while one of them (gloss) can be measured very accurately, there is a large inherent uncertainty in the measurement of the other (drying time), constituting thus an interesting test-bed for ANN's. The set of variables that most significantly affect one of these given properties can also be found through a sensitivity analysis of the response hypersurface on the input variables. These are shown to mostly coincide with the expected chemical and/or physical influence of the formulation components in the case of gloss.

Fundamentals

An "artificial neuron", the basic element of an ANN, receives various stimuli (cf. input from "dendrites") $\{x_i\}_{i=1,...,n}$, and gives a certain response (cf. "axion potential") $y = f(x_i)$. Below a certain threshold: $y = 0$. Above a certain value of the input, the neuron gets saturated: $y = 1$. For one input x, this can be represented by $y = tanh(x)$ (which is convenient as hyperbolic tangents form a complete set for continuous functions in [0,1]). It is customary to use instead "sigmoidal" functions (a simpler form of $tanh(x)$)

$$y = f(x) = \frac{1}{1 + e^{-a(x-T)}} \quad , \quad (1)$$

where T is the inflexion point, and a is a measure of how fast y grows from near $y = 0$ to near $y = 1$. Sigmoidal functions (in particular) can approximate any continuous (actually Riemann integrable) function which is bounded in compact sets (cf. *9, 10*). For many inputs, one can take

$$x = \sum_i w_i x_i \quad (2)$$

where w_i are weights to be adjusted depending on the system to be studied. A schematic representation is shown in Figure 1 below. Adjusting the weights w is precisely what constitutes the "learning" of the ANN, and one achieves that by using a large database of known inputs and responses, which can be taken as a representative sample of the population in question (e.g. alkyd enamels of various types).

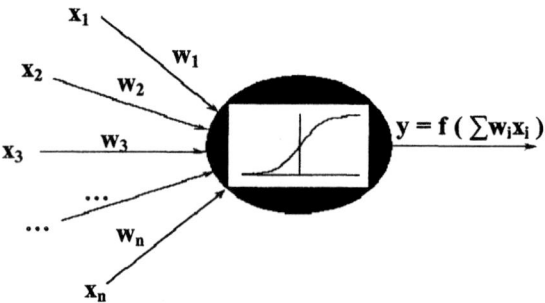

Figure 1. Schematic representation of the response of a neuron in a network. The inputs are usually the outputs of previous neurons, and are weighed by {w's}. The output comes from applying the "transfer" function f to the weighted sum of inputs.

The real power of a neural network comes from the following: Not all basic elements of the prepositional calculus can be expressed in terms of the simplest artificial neuron, but they can all be in terms of *combinations* of these neurons. It follows, then, that *any* mathematical expression can be written as a sequence of artificial neurons properly combined ("connected"). In this last statement, together with the approximation capabilities of sigmoidal functions referred to above, relies the potential applicability of ANN's to a whole plethora of systems regardless of their nature. Networks can be constructed by arranging neurons in "layers", the first of which is set by the input variables and the last by the output variables. The values are processed sequentially through the network depending on how the connections between neurons are made (cf. Figure 2). How one connects the neurons to each other constitutes the "art" of using ANN's: non-linear systems will normally demand more than one layer of neurons (where one layer usually receives inputs from the layer preceding it, and sends its outputs to the layer that follows; feedback mechanisms will require that some neurons have outputs connected to the inputs of neurons on previous layers; etc. Thus, the model under study can be changed if we change the connections between neurons (the "network architecture"), because we are changing which neurons have influence over a given one; and the model can be adjusted to the system by changing $\{w_i\}_{i=1, ..., n}$, because we are deciding how large is the impact of a neuron on another. Perhaps the main problem is, how can we construct the sequence of neurons *without* knowing the mathematical expression that we want to model?

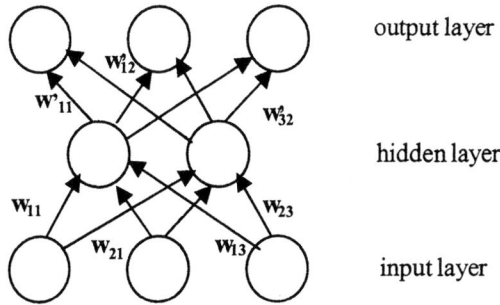

Figure 2. Simple multi-layered network. The intermediate layers between the input and output layers are usually called "hidden" layers (only one shown).

In this work we will use "feed-forward ANN's": communication exists only between neurons in adjacent layers, implemented through unidirectional connections ("weights"). The typical problem to be solved by an ANN is: given an unknown input element $(x_1, ..., x_n)$, to which class or group $(y_1, ..., y_m)$ does it belong? To this end, the network has to be *trained* first (to recognize the associations between them, using an available data set), and then used as a *predictive* tool (to yield the class to which a previously unseen input belongs).

Training

Training a network means finding an appropriate set of weights $\{w_i\}_{i=1, ..., n}$, for which the difference between true and predicted values is a minimum. Both the inputs *and* outputs are simultaneously given to the network and the w_i's are found through iteration as follows:

A set $\{w_i\}_{i=1,...,n}$ is first chosen randomly. If y_k represents the output of the ANN and Y_k is the true value, we define the $k'th$ error to be

$$E_k(w) = 1/2 \, |Y_k - y_k|^2 \qquad (3)$$

and the total error

$$E(w) = \sum_k E_k(w) \qquad (4)$$

We can calculate $\nabla E(w)$ and move $w \to w + \Delta w$ in the direction in which $E(w)$ decreases most. And so we iterate, until $E(w)$ reaches a value equal or less than a given tolerance. One way to update w at every iteration step j is by calculating the conjugate directions

$$\mu_j = -g_j + \frac{g_j^T g_j}{g_{j-1}^T g_{j-1}} \mu_{j-1} \qquad \text{(for } j > 1\text{)} \qquad (5)$$

where $g_j = \nabla E(w_j)$, and obtain the η_j that minimizes $E(w_j + \eta_j \mu_j)$.

Other algorithms make use of learning parameters and/or momentum terms. Regardless of the technique used to reduce the error systematically, the procedure leads to what is called "training the network". **A successful training consists of determining a set of weights** $\{w_i\}_{i=1,...,n}$ for which the total error in eq.(4) is a *global* minimum. As in all extremization problems, one can only guarantee *local* maxima or minima. Other methods such as the so-called "genetic algorithms" can be combined with ANN's in order to find consecutively lower minima, but these will not be treated here.

Over-Training

Great care should be taken with not over-fitting the model (cf. e.g. *11*). If $h(x)$ is the "real" unknown model that we want to approximate, and $y'(x)$ are the measured values, then, as all measurements carry an inherent uncertainty,

$$y'(x) = h(x) + \varepsilon(x) \qquad (6)$$

Therefore, if $y'(x)$ is adjusted by "over-training" the network, we will be adjusting the noise $\varepsilon(x)$ associated to the data (if only because the network has no information whatsoever on the true model $h(x)$). A simple way to prevent this over-fitting is dividing the data into two disjoint sets $T = \{\text{training data}\}$, $V = \{\text{validation data}\}$, such that $T \cap V = \emptyset$. Train the network using T and stop it randomly, testing it with V (not changing the w's in this step). As the ANN learns to model $h(x)$, the error in predicting V will diminish. But as the ANN starts learning the noise $\varepsilon(x)$ inherent in T, predictions on V will be poorer and one will know heuristically when to stop training.

Other (more complicated) methods for controlling over-training exist, based on controlling the curvature of $y'(x)$; this amounts, essentially, to reduce the amount of free parameters to a minimum. Whereas a model $y'_1(x)$ with too many free parameters will adjust the training data very well but may fail to predict $h(x)$ for new values of x, a model $y'_2(x)$ too poor in free parameters, while simpler, might not even adjust the training data well. Care should therefore be taken and there are no hard rules to follow.

Prediction

Here, $\{x_i\}_{i=1, ..., n}$ and $\{w_i\}_{i=1, ..., n}$ (the latter found through training) are given as inputs, and y_k is the output of the network (cf. Fig. 3). I.e., the network associates $\{x_i\}_{i=1, ..., n}$ to a class y_k.

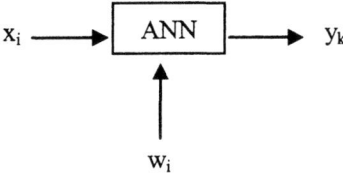

Fig. 3. Network used for prediction

The efficiency of the ANN is calculated as

$$\eta = \frac{\|\{correct\ associations\}\|}{\|\{all\ associations\}\|} \qquad (7)$$

Though no fundamental understanding of the dependence between the input and output variables is to be had, one is allowed to study complex systems for which a mathematical model or computational algorithm does not exist. Not only are the answers obtained in a very reasonable amount of time, but as the ANN's learn with more and more examples the efficiency increases.

Methodology and Application

We chose to work with alkyd enamels, due to their high volumes produced by industry and because they are simpler than latex paints. Five different families were chosen, viz. conventional enamels, high-solids enamels, modified (styrenated, acrylated, etc.) enamels, interpenetrating networks (made from mixtures of high-solids alkyd resins and acrylic resins), and enamels with reactive diluents. There are many variables of study, but, for mathematical simplicity, we chose to develop one network for each dependent variable. 57 independent variables were initially chosen (21 for the alkyd resin, 26 for the enamel formulation, 6 for the solvent, 1 process variable, and 3 redundant variables); and 2 output (dependent) variables: specular gloss at 60°, which ranged from matt to high gloss (and which can be very accurately measured), and drying time to touch, which ranged from a few minutes to over a day (and whose measurement carries a large uncertainty). All enamels (346 formulations in total) were formulated in the laboratory under controlled conditions. The breakdown of the database is given in Table I below. Of these, 13 formulations were removed for the gloss-ANN (as 12 were transparent paints and 1 had no data available). All were used in the study of drying time. The formulae for building the Training and Prediction Sets were chosen following the guidelines given in ref. (*1*); care was taken that those in the Training Set represented the full 5 populations, and about 2/3 of each.

Table I. Breakdown and distribution of enamel formulations used.

Type of Enamel	*Training Set*	*Prediction Set*	*Total*
conventional	69	40	109
high solids	110	72	182
reactive diluents	19	5	24
interpenet. ntwks.	4	2	6
modified	17	8	25

All values of independent variables $\{x_i\}_{i=1,...,57}$ were normalized in $[0,1]$. This is important because the values of different components of the input may differ by several orders of magnitude. The values of the dependent variable, y_k, were normalized in $[0.1, 0.9]$. The topology of the network was chosen to have one hidden (internal) layer with number of nodes n to be determined, and with all nodes in one layer connected to all the nodes in the next neighboring layer. The ANN used is referred to as a *57-n-1* network, of the "feed-forward" type, and the algorithm used to minimize the error was that of conjugate gradients. All weights $\{w\}$ were initialized pseudo-randomly in $[-0.05, 0.05]$.

Choosing an appropriate convergence tolerance $\varepsilon(x)$ (cf. eq. 6) is important, as mentioned above. In the case of gloss-60^0, the measurement error of the instrument is negligible but the human eye can hardly distinguish gloss values within a range of ±3, in a scale from 0 to 100. In contrast, the errors incurred in the measurements of drying times are at least around 25%, mainly because of ambient and human conditions that introduce subjective errors, in addition to the lack of a standardized measurement methods among different laboratories. These values were thus respectively chosen for $\varepsilon(x)$.

Determination of *n*

The minimum topology of the network is determined using the entire database, after proper normalization is performed. Too many hidden nodes mean too many free parameters, which in turn imply that the network could "learn" all correlations with no predictive power. Too few hidden nodes ask for a large number of iterations, which may not converge during the training phase. We tried to reproduce all the outputs from the inputs; for gloss, we attained good convergence for 6 nodes after 10 cycles (a *cycle* is a number of iterations equal to the number of weights), and poor convergence for 5 nodes; for drying time we had good convergence for 6 nodes after only 4 cycles, but convergence problems with 4 nodes. We thus chose *n=6* (network 57-6-1*)*.

Results and Analysis of Results

Using the trained network as described in previous paragraphs, and testing it with the Prediction Set, we are able to predict about 90% of the new formulae to within the chosen tolerance:

$$\eta_{gloss} = 90.16\% \ (\varepsilon=0.03); \quad \eta_{drying\ time} = 88.98\% \ (\varepsilon=0.25) \quad (8)$$

It is interesting to see that many predictions turn up to be very close to, but above of $\varepsilon(x)$, thus being discarded and accounted for as badly predicted; i.e., slightly increasing $\varepsilon(x)$ has a large effect on η. It is also interesting to note that several different seeds for initially randomizing the weights $\{w\}$ produce very similar prediction efficiencies, giving some robustness to the methodology.

The global result shown by eq.(8) can be analyzed by enamel family (cf. Table II). We see that modified enamels have the poorer predictability; although their number is small, the same can be said of other families.

Table II. Predicting power by enamel family.

Type of Enamel	η_{gloss} (%)	$\eta_{drying\ time}$ (%)
conventional	94.73	90.0
high solids	93.85	87.5
reactive diluents	75.00	100.0
interpenet. ntwks.	100.00	100.0
modified	55.55	87.5

Sensitivity Analysis

Complex systems do not lend themselves easily to analytical study. Having many measurable properties, very few known analytical dependencies, and several of the measurable properties being a function not only of the measuring method (!) but of a plethora of other (possibly hidden) variables, make mathematical models that describe their behavior unavailable.

However, as ∇y is a good linear estimate of how sensitive y is to each of the input variables, one could use the hypersurface $y(x)$ constructed by the ANN to predict expected variations in a property value of a given paint formulation due to a small variation in the amount of a certain component: stimuli–response via a sensitivity analysis. One expects ∇y to be similar when evaluated at formulations within the same family (type of enamel), but would differ substantially from family to family, as other parameters come into play more strongly when the chemistry of the system is changed.

It is important to avoid spurious variations due solely to uncertainties in the measured inputs. To this end, if $y_k^{(t)}$ represents the value of the property k of a formulation belonging to a given enamel family of type t, then the sensitivity of this property to a variation in a particular x_i can be defined by

$$S_{ki}^{(t)} = F_i^{(t)eff} \frac{\partial y_k^{(t)}}{\partial x_i} \qquad (9a)$$

where $F_i^{(t)eff}$ is an *effectivity parameter* that can take only two values (*0* or *1*) according to the percentage variation of the component x_i with respect to its mean value $x_i^{(t)}$ within the family:

$$F_i^{(t)eff} = \begin{array}{ll} 1 & \text{if } \Delta x_i^{(t)}/x_i^{(t)} > 0.1 \\ 0 & \text{if } \Delta x_i^{(t)}/x_i^{(t)} < 0.1 \end{array} \qquad (9b)$$

Here, $\Delta x_i^{(t)} = x_i^{(t)max} - x_i^{(t)min}$. Then $S_{ki}^{(t)} > 0$ means that the property y_k increases whenever x_i increases, and $S_{ki}^{(t)} < 0$ means it decreases.

∇y was then calculated and evaluated within the different enamel families. For comparison purposes, all values were normalized with respect to the largest derivative in absolute value. Also, in order to avoid false variations due to surface fitting, a smoothing of the response hypersurface was carried out before calculating the gradients (cf. 1). This allows the network to learn input-output relationships without the particular details of each formulation. The results for y_k = gloss are shown in Table III below; up to 6 variables are shown in order of importance (more were calculated). The hierarchical order of importance of the inputs is maintained throughout each family. We can see that they are in very good consistency with our experience in all cases. The table is then a good guideline for enamel reformulation: if we are interested, for instance, in changing the gloss of an enamel formulated with reactive diluents, our best bet is to change the amount of slow solvent or the PVC; if other constraints do not allow us to do that, we should change the amount of zirconium; etc., and similarly for the other families.

Unfortunately, the same cannot be obtained for drying time. While some of the most significant variables obtained from the sensitivity analysis are consistent with what is expected, some others are not well understood. Although the network constructed offers very good predicting power, the large variations in the determination of this property prevents the "smoothed" network from a satisfactory analytical description of its behavior. Note that predictive power is important even at 25% uncertainty here: an enamel that should dry in 7 min., say, will not be discarded if it dries in 9 min., so as an enamel that is expected to dry in 4 hrs. if it does in 5 hrs. But derivatives are, in this sense, much more sensitive.

Concluding Remarks

Intelligent systems in the form of artificial neural networks can be constructed successfully as a tool for predicting and analyzing coating properties. In the case of alkyd enamels of several different types, a predicting power of the order of 90% can be attained. Through a sensitivity analysis of the parameters under study, ANN's can provide us with the set of most significant variables that affect a given property in a complex system. The response of the system to a given change in inputs may thus not only be predicted, but studied analytically to a first approximation. This in itself constitutes a powerful tool for reformulation.

Table III. Sensitivity analysis for *gloss*.

Enamel Family	Significant Variables x_i (in order of importance)	Sign of $\partial y/\partial x_i$	Consistency w/experience
Conventional Enamels	matting agent	< 0	yes
	oil absorption	< 0	yes
	slow solvent	> 0	yes
	PVC	< 0	yes
	emulsifying agent	< 0	yes
	oil length	> 0	yes
High-Solids Enamels	matting agent	< 0	yes
	oil absorption	< 0	yes
	slow solvent	> 0	yes
	Hegmann fineness	> 0	yes
	PVC	< 0	yes
	emulsifying agent	< 0	yes
Enamels w/ Reactive Diluents	slow solvent	> 0	yes
	PVC	< 0	yes
	p/b	< 0	yes
	drying agent	< 0	yes
	alkyd content	> 0	yes
	fast solvent	< 0	yes
Enamels from Interpenetrating Networks	slow solvent	> 0	yes
	PVC	< 0	yes
	oil length	> 0	yes
	p/b	< 0	yes
	drying agent	< 0	yes
	alkyd content	> 0	yes
Modified Enamels	matting agent	< 0	yes
	oil absorption	< 0	yes
	slow solvent	> 0	yes
	Hegmann fineness	> 0	yes
	PVC	< 0	yes

Future Work

Other output variables need to be studied. It is obvious that the network that best describes one variable will not in general be good to describe another variable, as the model $y_k = f_k (x_1, ..., x_n)$ is specific for y_k. However, careful handling should allow one to construct an ANN with vectorial output which could handle all paint properties at once without necessarily sacrificing efficiency.

In order to improve on error minimization a more powerful model could be constructed by linking ANN's with so called genetic algorithms. These are randomized parallel search methods that pretend to simulate the natural evolution process, and have proven to be very useful.

Newer probabilistic methods such as *Probabilistic Neural Networks* (PNN) and *Generalized Regression Neural Networks* (GRNN) could also be contemplated; these are being used both to classify and for quantitative structure-property relationship studies applicable to continuous function mapping problems (*12*).

As in all works using artificial neural networks, the sample population needs to be increased. Most literature report samples of $\mathcal{O}(10^3)$, while we are in the low $\mathcal{O}(10^2)$. Especially in the case of those families for which the number of samples is very low, and in that one for which the prediction is poor, a greater coverage of parameter space would be desirable.

Acknowledgements

The authors are indebted to Adela Reyes and René Nakamura for useful discussions and feedback, to Héctor Hernández for training and running the networks, and to Martín Alvarado for making new enamel formulations.

References

1. Vitela, J.E.; Nahmad-Achar, E. *Eur. Coat. J.* **2004**, to appear.
2. Watanabe, K.; Matsura, I.; Abe, M.; Kubota, M.; Himmelblau, D.M. *AIChE J.* **1989**, *35*, 1803-1812.
3. Bhat, N.V.; Miderman Jr, P.A.; McAvoy, T.; Wang, N.S. *Modeling Chemical Process Systems via neural Computation;* IEEE Control Systems, 1990; Vol. 10 No. 3.
4. McAnany, D.E. *ISA Trans.* **1993**, *32*, 333-337.
5. Ramesh, K.; Tock, R.W.; Narayan, R.S.; Bartsch, R.A. *Ind. Eng. Chem. Res.* **1995**, *34*, 3974-3980.

6. Tsen, A.Y.; Jang, S.S.; Wong, D.S.H.; Joseph, B. *AIChE J.* **1996**, *42*, 455-465.
7. Dhib, R.; Hyson, W. *Polym. React. Eng.* **1997**, *10*, 101-113.
8. Chen, L.; Gasteiger, J. *J. Am. Chem. Soc.* **1997**, *119*, 4033.
9. Leshno, M.; Lin, V.Y.; Pinkus, A; Schocken, S. *Neural Networks* **1993**, *6*, 861-867.
10. Hornik, K. *Neural Networks* **1991**, *4*, 251-257.
11. A very good explanation on this topic can be found in Schraudolph, N. and Cummings, F. *Introduction to Neural Networks*, lecture notes given at the Istituto Dalle Molle di Studi sull'Intelligenza Artificiale, Lugano, Switzerland, 2001.
12. Mosier, P.D.; Jurs, P.C. *J. Chem. Inf. Comput. Sci.* **2002**, *42*, 1460-1470.

Chapter 12

Maleimide–Vinyl Ether-Based Polymer Dispersed Liquid Crystals

Molly Hladik[1], Askym F. Senyurt[1], Charles E. Hoyle[1], Joe B. Whitehead[2], Charles M. Werneth[2], and Garfield T. Warren[2]

[1]School of Polymers and High Performance Materials and [2]Department of Physics and Astronomy, The University of Southern Mississippi, Hattiesburg, MS 39402

A photocurable crosslinked matrix forming mixture of bismaleimide and divinyl ether monomers has been used to produce polymer dispersed liquid crystals (PDLCs) with fast optical switching times and excellent transmission characteristics. The use of a divinyl ether with a urethane space group resulted in effective scattering of light in the off state and reasonably high transmission in the on state. The resin required no addition of an external initiator; consequently the PDLCs are relatively stable to artificial weathering. The maleimide/vinyl ether resins systems described herein have the potential for serving as non-yellowing, rapid cure resin matrices for photocurable PDLCs.

1. Introduction

Polymer dispersed liquid crystals (PDLCs), fabricated by a number of phase separation methods to achieve a dispersion of micron size droplets of a low molecular weight liquid crystal or liquid crystalline mixture in a solid polymer matrix [1-3], operate by electrically modifying the ordinary refractive index of the liquid crystal droplets to match the refractive index of the polymer matrix [4]. Optimum operating parameters for PDLCs include low switching voltages, short switching times, high contrast ratios, and thermal/photostability. The applications for PDLCs range from light shutters to automobile displays, flat panel displays, and high resolution projection displays [5,6].

The most popular method of making PDLCs is by polymerization-induced phase separation wherein small molecule photoinitiators are used to initiate the polymerization of a combination of polymerizable monomers and non-polymerizable liquid crystal molecules. The liquid crystal component phase separates upon polymerization due to solubility and elastic driving forces. Polymerization of traditional monomer systems (either thiol-enes or acrylates) to produce the continuous polymer network from which liquid crystal separate are initiated with about 1-5 weight percent of a conventional photoinitiator. After the polymerization is complete, a significant quantity of the photoinitiator is still present in the matrix. This may result in alteration of the phase transitions in the liquid crystalline phase by virtue of the photoinitiator being dissolved in the liquid crystalline phase. The residual photoinitiator and its photo by-products may also cause degradation of the PDLC film upon further exposed to light.

The systems described herein use maleimides as both initiator and comonomer, thus eliminating the need for an external photoinitiator. Maleimides have potential uses in PDLC systems because of their photo-reactivity (and photostability after polymerization) which allows them to initiate polymerization without the addition of a traditional photoinitiator when exposed to ultraviolet (UV) light [7-10]. Since the maleimide functional group is also incorporated into the polymer matrix as a comonomer as has been reported in a large number of past publications [references 7-10 are representative examples], there are little or no residual components in the matrices formed by exposure of bismaleimide/divinyl ethers to light. Such maleimide/vinyl ether systems have already been shown to offer potential as photocurable clear coats in a variety of applications, both due to the final properties afforded to the cured films as well as their ability to achieve unprecedented rates of polymerization under ambient conditions in air. The PDLCs described in this paper are comprised of a bismaleimide mixed with two vinyl ethers, one with a urethane spacer and the other with an ester spacer. This work clearly demonstrates that the urethane group is essential to achieving maleimide/vinyl ether based PDLC networks with acceptable electro-optic performance characteristics.

2. Experimental

2.1. Materials and Sample Preparation

Q-BOND, a commercial bis-maleimide with the structure shown below, was obtained from Loctite Corporation. The VEctomer samples were obtained from Allied Signal and have molecular weights of 436 g/mole (Vectomer 4020) and ~4,300 g/mole (Vectomer 2020). The thiol-ene system from Norland Optical Adhesives (Norland 65) has been described [11]. The liquid crystal mixture E7 is a mixture of four cyano multiphenyls (structures shown below) in the following percentages: **K15**-51%, **K21**-25%, **M24**-16% and **T15**-8%. The transition from the nematic phase to the isotropic phase (T_{NI}) for pure E7 is 61 °C.

Pre-polymer bis-maleimide/divinyl ether mixtures used to fabricate PDLC cells were comprised of 60 mole % Q-BOND and 40 mole % of VEctomer 4020 and VEctomer 2020 mixtures. The VEctomer 4020/2020 mixtures were prepared by varying the percentage of VEctomer 2020 from 0.49 mole % to 7.6 mole % (this corresponds to weight percent of VEctomer 2020 ranging from 10 to 70 weight %) as shown in Table 1. Each pre-polymer sample was mixed with 60 weight % liquid crystal E7 to prepare the formulations used to make the PDLC samples. The PDLC cells were prepared by capillary filling the liquid crystal/pre-polymer mixture into transparent indium-tin oxide (ITO) coated glass substrates. ITO coated glass cells were constructed by using 10 µ glass fiber spacers as shown in Figure 1: we note that due to shrinkage and other considerations, the thickness of the final cured PDLCs was less than 10 µ. Prior to UV light exposure, the capillary filled liquid crystal, pre-polymer cells were allowed to equilibrate for an hour before exposing to a 450 W medium pressure mercury lamp with a light intensity of ~30 mW/cm^2 for 5 minutes to make the cured PDLC cell..

2.2. Photodifferential Calorimetry (Photo-DSC)

A photo-DSC was used to obtain the polymerization rate and percent conversion of each photocurable PDLC sample. The maximum peak height of the photo-DSC curve is proportional to the polymerization rate while the area under the curve is proportional to the total monomer conversion. A modified Perkin-Elmer DSC-7 was used to record the exotherms for the monomer mixtures. In a crimped, cleaned and dried aluminum DSC pan, 2 µL of the monomer mixture was dispensed with a calibrated microsyringe. The samples were then placed in the DSC and purged with nitrogen for 1 minute. The light source used to irradiate the samples was a 450 W medium pressure mercury lamp from Canrad Hanovia (~30 mW/cm^2). The light passed through quartz before reaching the sample and reference cells. Sample compositions for photo-DSC and refractive index measurements were approximately the same as used in the PDLC fabrication.

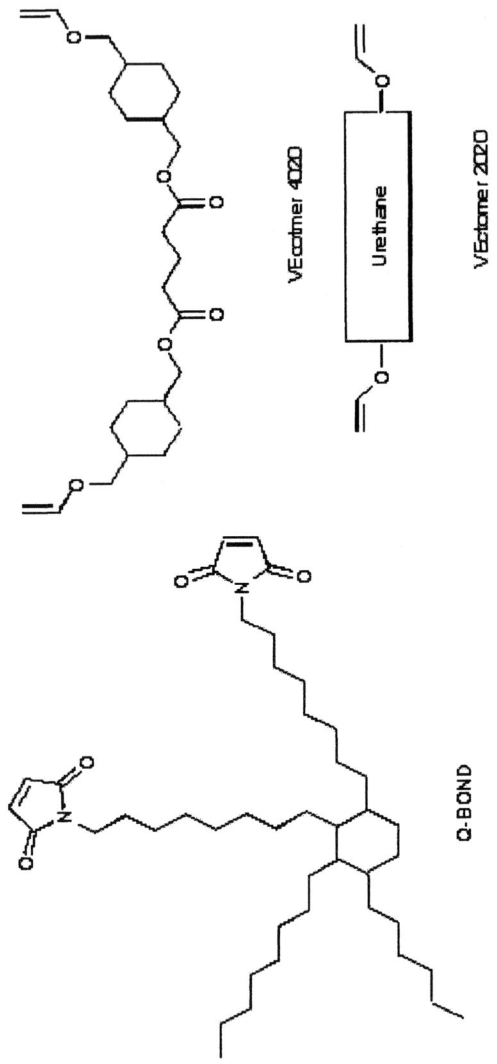

VEctomer 4010

VEctomer 2020 Urethane

Q-BOND

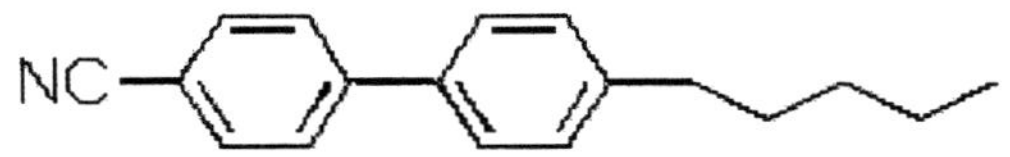

Chart 1. Structures of monomers and liquid crystals comprising E7.

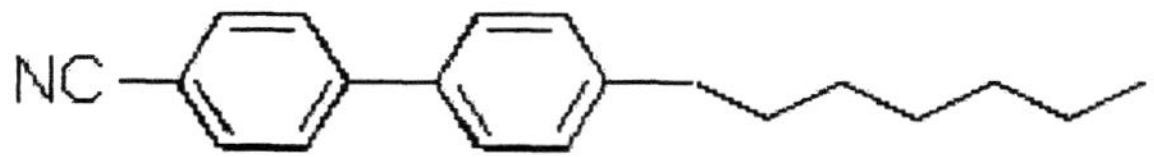

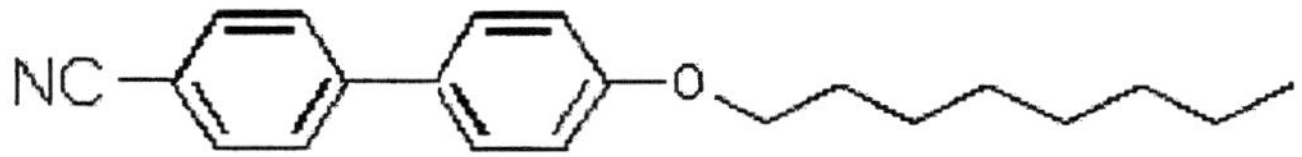

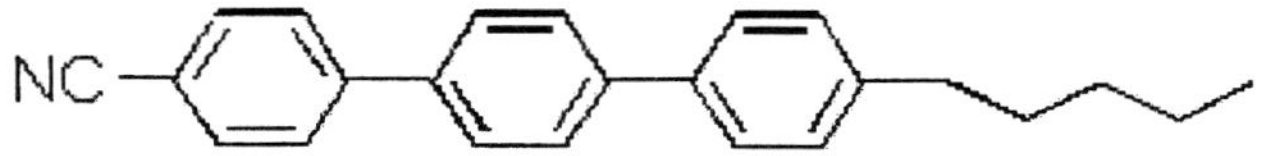

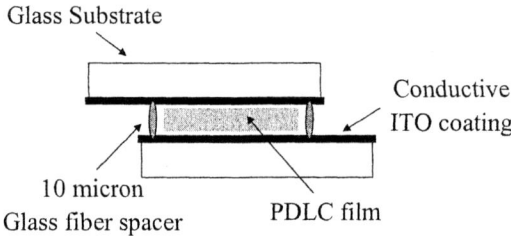

Figure 1. Schematic Representation of PDLC cell

2.3. Electro-Optic Measurements

Figure 2 is an overall diagram of the optics bench and equipment used to collect and process the electro-optic switching characteristics of each PDLC cell. To make the electro-optic measurements, the requisite cured PDLC cell was placed in the sample holder and alligator clips were placed on either side. The alligator clips were connected to the power supply and the function generator, which generates a ramped sinusoidal waveform output from the computer. The voltage amplitude, V_{90}, required to achieve ninety percent of the difference between the maximum percent transmission and the minimum percent transmission is then applied to the PDLC cell in the form of a sinusoidal pulse waveform (note that V_{90} is an rms voltage amplitude of a sinusoidal waveform). From the resultant percent conversion versus time plots the percent transmissions in the off (T_{off}) and on (T_{on}) states are obtained. The resultant Contrast is calculated from the ratio T_{on}/T_{off}. The switching time, τ_{switch}, is the sum of the turn-on time, τ_{on}, defined as the time required upon application of V_{90} to go from 10 to 90 percent of the difference in T_{on} and T_{off}, and the turn-off time, τ_{off}, defined as the time to go from 90 to 10 percent of the difference in T_{on} and T_{off} after the voltage is turned off.

2.4. Refractive Index Measurements

Refractive index measurements of cured films and pre-polymer mixtures of liquid monomers were made with a standard refractometer.

2.5. Scanning Electron Microscopy (SEM)

An environmental scanning electron microscope (FEI, Quanta 200) was used to examine the phase separated morphology of PDLC cells. The samples, after electro-ooptic measurements were made, were freeze-fractured, soaked in hexanes for 24 hours, and placed under vacuum for 12 hours before examination by the electron microscope. The freeze-fractured samples were mounted on the sample holder with double side tape and sputter coated with gold (Polaron E5100 sputter coater) in order to enhance the quality of the images.

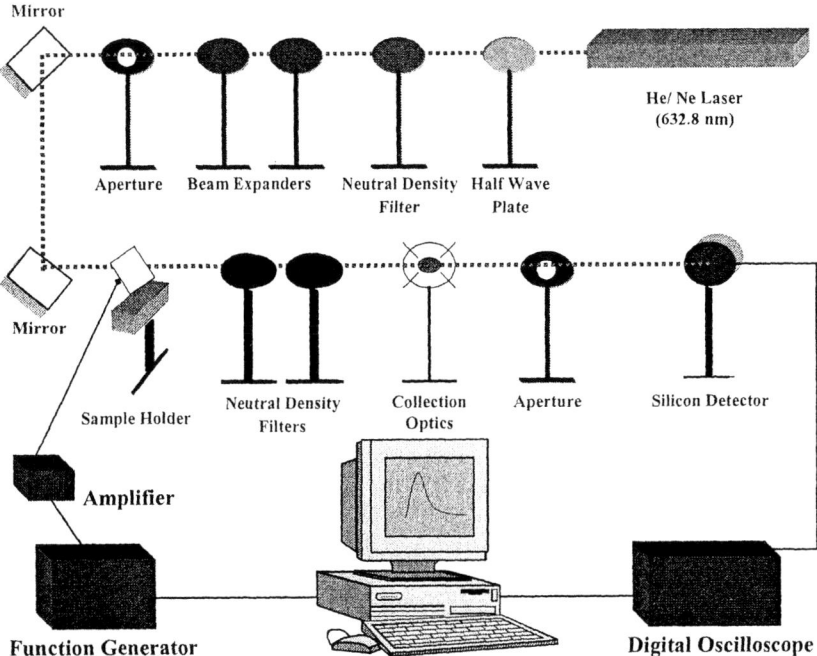

Figure 2. Diagram of set up for electro–optic measurements of PDLC cells.

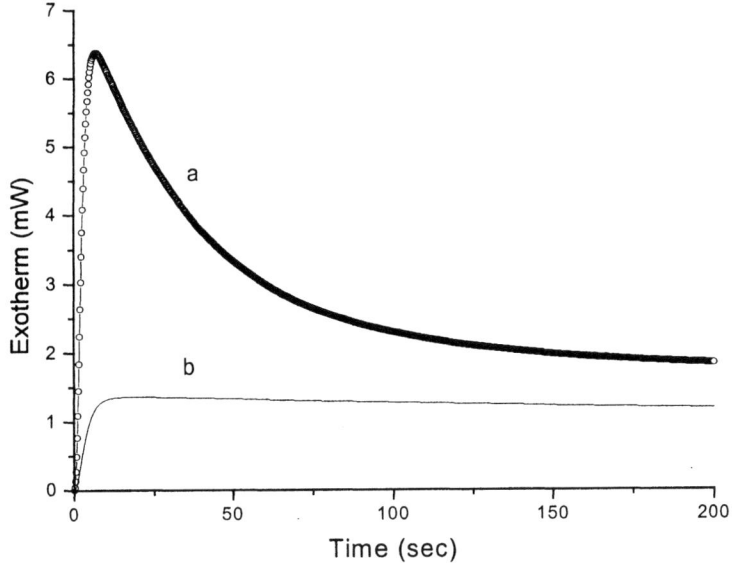

Figure 3. Photo-DSC for (a) Q-BOND and (b) Vectomer 4020 with no added photoinitiator. (Light intensity ~ 29.3 mW/cm^2).

3. Results and Discussion

This section presents the results for PDLCs based upon divinyl ethers and a bismaleimide monomer combined with the liquid crystal mixture E7 as described in detail in the Experimental section and given in Table 1. Polymerization exotherm curves for several maleimide/vinyl ether systems without liquid crystal present were first recorded. Refractive index measurements were made before and after polymerization for the mixtures that were subsequently combined with E7 and polymerized to give the PDLCs. Electro-optic measurements, which include switching voltages (V_{90}), the percent transmittances and Contrasts in the on-off states (T_{on}, T_{off}) upon application of V_{90}, and switching times (τ_{switch}) are also reported for each PDLC network made (see Experimental section for definition of electro-optic parameters). The results for the maleimide/vinyl ether based PDLCs are compared with results for a conventional thiol-ene system photocured under identical conditions.

Table 1. Pre-polymer mixture formulations.

Sample	Mole % Q-bond	Mole % VEctomer 4020	Mole % VEctomer 2020
1	59.81	39.70	0.49
2	59.65	39.35	0.99
3	59.74	38.53	1.73
4	59.65	37.73	2.62
5	59.54	36.74	3.72
6	59.87	34.83	5.29
7	60.06	32.34	7.60

3.1. Photo-DSC Exotherms for Maleimide/Vinyl Ether Polymerization

First, as a reference, polymerization results were obtained for the bismaleimide and the pure low molecular weight divinyl ether Vectomer 4020. The results in Figure 3 for the photoinduced homopolymerization of pure vinyl ether and pure bismaleimide (Q-BOND) upon exposure of a medium pressure mercury light source clearly show that pure divinyl ether (VEctomer 4020) gives no exotherm while the bismaleimide readily homopolymerizes, i. e., the heat recorded for the Vectomer 4020 is strictly due to absorption of light and the resultant heat transfer to the heat sink as Vectomer 4020 does not polymerize upon exposure to the medium pressure mercury light.

Next, Q-BOND was irradiated in the photo-DSC with 60 weight percent of E7 added to determine the effect that adding the liquid crystal, which absorbs light effectively in the range where the medium pressure mercury light emits, has on the exotherm. By comparing the exotherm of the Q-BOND sample without (curve a) and with (curve b) the added E7 it is apparent that the exotherm, while lowered in the presence of the liquid crystal, still occurs representing polymerization of the maleimide upon absorption of light. The lowering of the exotherm compared to the case where no E7 is present is a result of a dilution effect leading to lower maleimide concentration and competitive absorption by the E7 liquid crystal as already mentioned. Similar results to those in Figure 4 have been obtained with bismaleimide/divinyl ether systems upon mixing with E7: this will be important in preparing PDLC cells of maleimide/vinyl ether/E7 mixture with no added photoinitiator

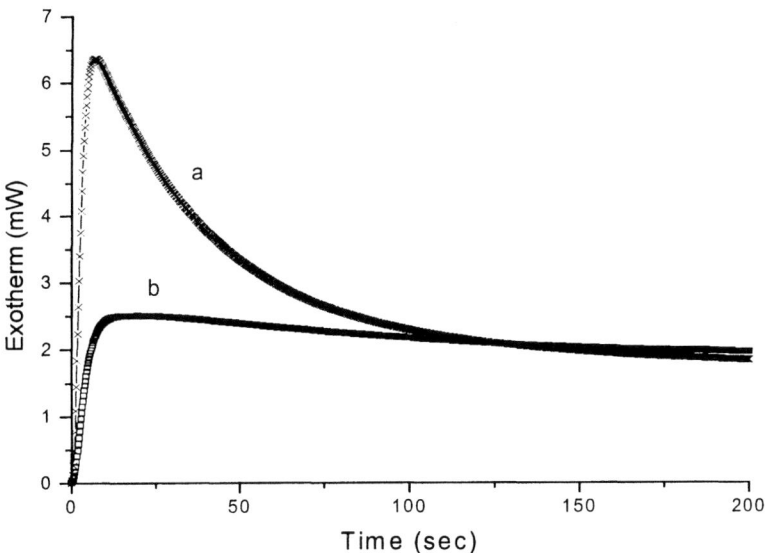

Figure 4. Photo-DSC for (a) Q-BOND and (b) Vectomer 4020 with no added photoinitiator. (Light intensity ~ 29.3 mW/cm^2).

Next, Q-BOND and VEctomer 4020 were mixed in increasing mole percentages of maleimide, and the copolymerization process was evaluated by photo-DSC. Figure 5 shows the polymerization exotherm curves for each mixture of Q-BOND and VEctomer 4020. From the exotherms in Figure 5, it is obvious that the rate of polymerization, as indicated by the higher exotherm, increases as vinyl ether is added. This is due to the faster polymerization rate for the copolymerization of maleimide and vinyl ether than the homopolymerization of the maleimide. The Photo-DSC exotherm peak-maxima (H_{max}) from Figure 5, which are proportional to the maximum rate of polymerization, are plotted in Figure 6. As found in earlier studies of the copolymerization of bismaleimides and divinyl ethers [8], the Q-BOND/VEctomer 4020 system attains a maximum exotherm rate for systems with ~50-60 mole % maleimide (Q-BOND). Note as we have already pointed out, the Vectomer 4020 sample with no Q-BOND present gives a very low heat flow that is essentially flat corresponding only to absorption of light leading to no polymerization of the vinyl ether groups.

In anticipation of the preparation of PDLCs from various maleimide/divinyl ether combinations, additional vinyl ether mixtures consisting of increasing concentration of VEctomer 2020 combined with VECtomer 4020 (divinyl ether with urethane oligomer spacer) were mixed with Q-BOND and the exotherms of the systems recorded in Figure 7. For now, we simply note that, as for the results in Figure 6, the maximum of the H_{max} versus bismaleimide mole percent plots for each sample occurs at about 60 mole % of maleimide functional groups . The excess of Q-BOND required to attain a maximum rate may result from some maleimide homopolymerization which occurs in addition to the maleimide/vinyl ether copolymerization. In view of the rapid rates achieved in the 60:40 molear mixtures of Q-BOND with total vinyl ether concentration, the pre-polymer PDLC mixtures used to prepare all the PDLC films which are the subject of the electro-optic measurements in this report are 60:40 maleimide/vinyl ether molar mixtures.

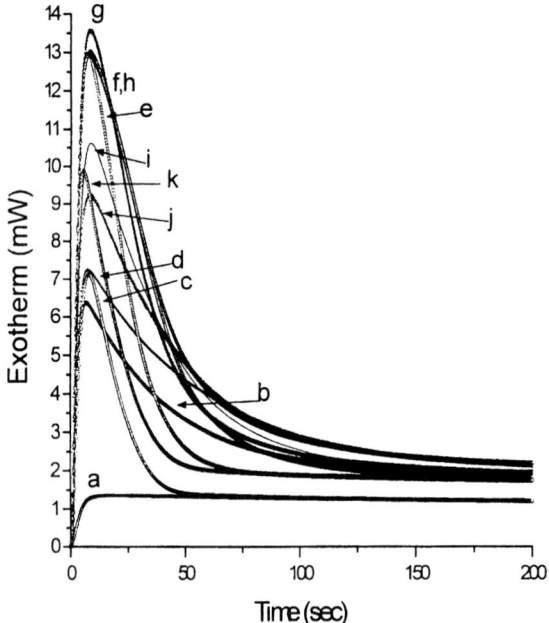

Figure 5. Photo-DSC exotherms for mixtures of Q-BOND and VEctomer 4020 ranging from 0 to 100 mole% Q-BOND: (a) 100% Vectomer 4020, (b) 100% Q-BOND, (c) 10% Q-BOND, (d) 20% Q-BOND, (e) 30 % Q-BOND, (f) 40 % Q-BOND, (g) 50% Q-BOND, (h) 60% Q-BOND, (i) 70% Q-BOND, (j) 80 % Q-BOND, (k) 90 % Q-BOND (Light intensity ~29.3 mW/cm^2).

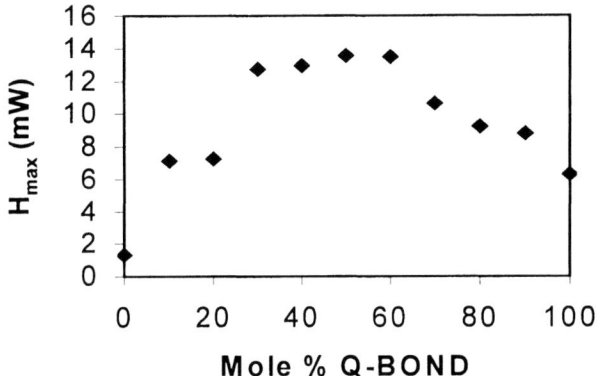

Figure 6. Plot of Hmax versus mole percent Q-BOND in mixtures of Q-BOND and VEctomer 4020. (Light intensity ~29.3 mW/cm^2).

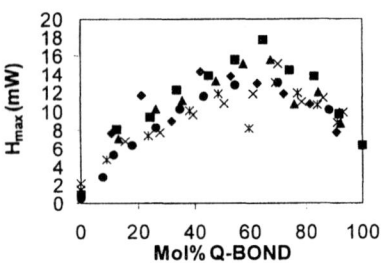

Figure 7. Exotherm maximum $|(H_{max})$ versus mole percent Q-BOND in mixtures of Q-BOND: VEctomer 4020: VEctomer 2020. All mixtures are composed of Q-BOND/ VEctomer 4020/Vectomer 2020 with the mole % of VEctomer 2020 as follows: (♦) 0.43 mole %, (■) 0.87 mole %, (7) 1.65 mole %, (□) 2.50 mole %, (□) 3.58 mole %, and (●) 5.16 mol %.

3.2. Refractive Index Changes of Bismaleimide/Divinyl Ether Based PDLCs

Before proceeding to report electro-optic measurement of PDLC films prepared from the maleimide/vinyl ethers in Table 1, results for refractive index measurements of sample films prepared by exposure to mercury light without bismaleimide/divinyl ethers systems are now presented. First, however, we will discuss the rationale for choosing the samples that were prepared for electro-optic measurements. Previously, we determined that switching voltages for PDLC films prepared from either Q-BOND alone, or mixtures of Q-BOND and VEctomer 4020, required switching voltages that were too high to allow accurate measurements. However, as we will demonstrate in this section, PDLCs prepared from Q-BOND and mixtures of VEctomer 4020 (low molecular weight divinyl ether) and VEctomer 2020 (a high molecular weight divinyl urethane), because of the presence of urethane groups which limit the solubility of the small molecule liquid crystals that comprise E7, exhibit lower switching voltages and excellent electro-optic switching properties. PDLC samples designated P1-P7 were made by preparing the pre-polymer monomer mixtures in Table 1 with 60 weight % E7, then making PDLC cells as described in the Experimental section, and subsequently exposing them to the 30 mW/cm^2 output of a medium pressure mercury lamp for 5 minutes. Switching results are shown in Table 2 for the PDLCs P1-P7. We begin by evaluating the switching voltages, V_{90}, required to switch the PDLC cells as defined previously. In general, the V_{90} values in Table 2 decrease as the Vectomer 2020 content increases, i.e., as the urethane content increases, indicating that samples with higher urethane content can be switched with lower voltages. Next we consider the actual electo-optic switching plots of percent transmittance versus time for two of the PDLCs: samples P1 and P7 shown in Figures 7 and 8, respectively.

Table 2. Electro-optic switching data for maleimide/divinyl ether PDLCs.

Sample	Mole% VEctomer 2020	τ_{switch} (msec)	V_{90} (Volts)	T_{off}	T_{on}	Contrast (T_{on}/T_{off})
P1	0.49	2.9	153	32.1	75.0	2.3
P2	0.99	3.0	157	27.6	73.0	2.6
P3	1.73	4.8	138	26.0	74.9	2.9
P4	2.62	9.2	12.3	5.0	76.8	15.3
P5	3.72	4.0	160	1.4	58.7	42.1
P6	5.29	7.6	108	2.7	78.5	29.3
P7	7.60	5.0	74	3.6	74.6	20.6

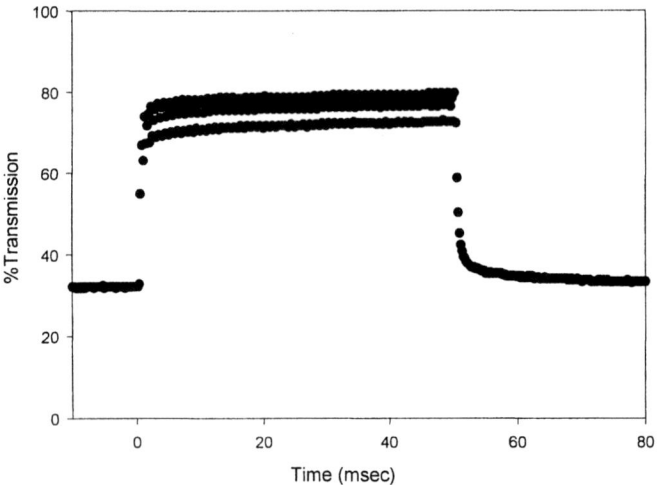

Figure 8. Electo-optic response of PDLC sample P1.

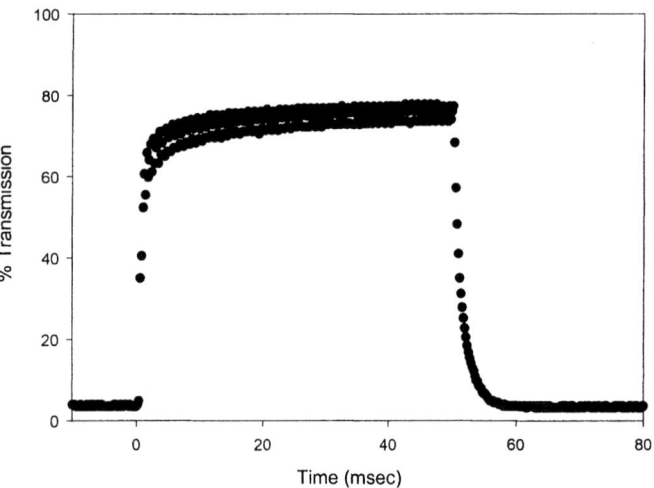

Figure 9. Electo-optic response of PDLC sample P7.

The results in Figures 8 and 9 are for PDLC samples with very different concentrations of urethane groups dictated by the concentration of VEctomer 2020 used in the pre-polymer formulation. As mentioned previously, the basic parameters used to determine the performance of the PDLC are related to optical transmission in the off and on states. The percent transmission in the off state, T_{off}, determines how effectively the PDLC is in blocking light transmission in the off state, while T_{on}, the percent transmission in the on state upon application of V_{90}, determines the clarity of the PDLC in the on state. The ratio of T_{on} to T_{off} is the Contrast of the PDLC cell. As seen in Figure 9, the P7 sample has a much lower T_{off} value than P1, and hence a significantly larger Contrast since both samples have approximately the same T_{on} values. The electro-optic switching values responses (including the switching times that have already been defined) for Figures 8 and 9 as well as data from all of the other PDLCs evaluated are collected in Table 2. It is obvious that as the amount of VEctomer 2020 incorporated into the PDLC increases, the switching time increases, the T_{off} value decreases, the Contrast value increases, and as already stated the V_{90} value decreases (note that the results for P5 are anomalous probably resulting from nonuniformity in the film). Interestingly, the T_{on} values are unaffected by the VEctomer 2020 content. The results in Table 2 thus demonstrate the importance of the urethane group in leading to PDLCs with enhanced electro-optic performance. i.e., low switching times, high Contrast, and low switching voltages. We speculate that the enthalpic interactions between the urethane backbone and the molecules comprising E7 are not favorable leading to enhanced phase separation driven by unfavorable free energy of mixing between urethane groups and liquid crystal molecules rather than the elastic free energy term since the molecular weight between crosslinks would increase with increasing VEctomer 2020 content.

3.4. Electra-optic Behavior of Thiol-Ene Based PDLCs

A PDLC from an example thiol-ene system with photoinitiator (see Experimental) was made to compare with the data presented for the bismaleimide/divinyl ether based PDLCs in the previous section. The PDLC made from the thiol-ene system with 60 weight % E7 under the conditions (sample polymerized with 30 mW/cm^2 medium pressure mercury light source) used to make the bismaleimide/divinyl ether based PDLCs had a V_{90} switching voltage of 29.7 volts and T_{on}, T_{off}, and Contrast values of 2.0, 67.5 and 33.8. The P7 PDLC in Table 2 has a faster switching time and higher value for T_{on} than for the thiol-ene PDLC, but a larger switching voltage and slightly higher T_{off} value. Overall the electro-optic properties of the bismaleimide/vinyl ether based PDLCs compare favorably with the thiol-ene system.

3.5. Microscopic analysis of PDLC films

In order to properly correlate the electro-optic data with respect to the chemical structure of the maleimide/vinyl ether based systems, it is essential to analyze the morphology of the cured PDLC films. Although we recorded the electron micrographs of each PDLC film, only the SEM micrographs of P1 and P7 are given in Figures 10 and 11 since they represent the extremes in electro-optic response. The micrograph in Figure 10 for P1 shows a morphology that is non-uniform across the thickness of the film with a series of small phase separated liquid crystalline droplets in the top half of the film and little or no droplets in the bottom half of the film. This may result from polymerization at the top and subsequent phase separation of liquid crystalline droplets which occurs early in the photolysis process due to the high absorbance of the maleimide groups in the top part of the film. As the top cures, the maleimides groups will react allowing deeper penetration of the light source and subsequent curing near the back surface of the cell. But, polymerization at the back of the cell does not result in the formation of liquid crystalline droplets. This may be due in part to depletion of liquid crystalline material in the lower half of the film as the curing and phase separation process in the top half takes place in the initial stages of the film formation process. Apparently, the presence of very small scattering particles in only the top half of the film results in relatively inefficient light scattering in the off state that gives a transmittance that is fairly high ($T_{off} \sim 32$ for P1). In the case of P7, much larger liquid crystalline droplets are found in the center of the film. The larger droplets (with sizes much greater than the wavelength of light) form at the expense of smaller droplets near the surface. The overall result is an enhancement in the light scattered in the off state: there is a longer path length contributing to light scattering. Of course, a consequence of a significant concentration of larger droplets is a rapid switching time and small switching voltage, as noted in Table 2 for P7. Although we did not include SEM micrographs for each of the other PDLC films reported on in Table 2, we note that there appears to be a decrease in the formation of smaller liquid crystalline droplets near the top surface along with the concomitant increase in the formation

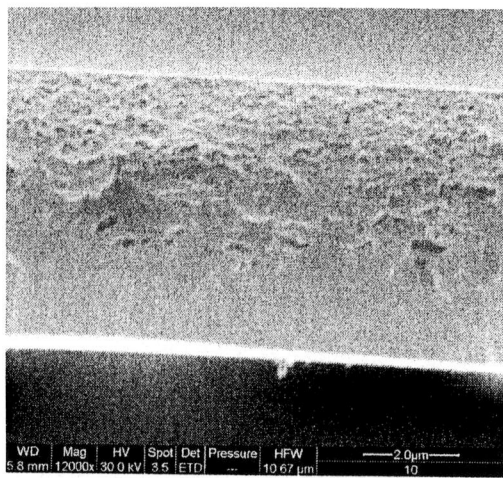

Figure 10. SEM of P1 PDLC film.

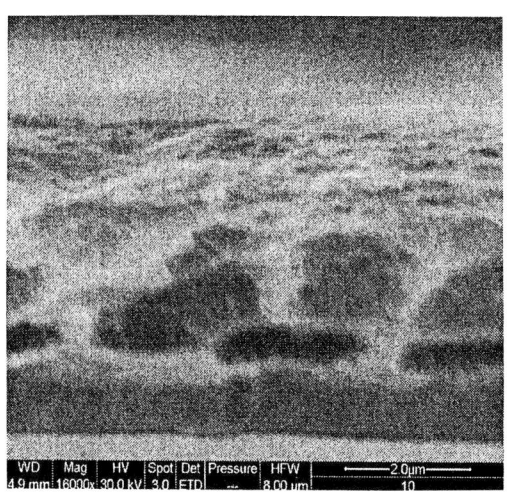

Figure 11. SEM of P7 PDLC film.

of larger droplets that span a greater distance across the PDLC film as the concentration of VEctomer 2020 increased.

3.6. PDLC long term light exposure evaluation

A preliminary evaluation was performed to determine the extent of selected PDLC film yellowing when exposed to an external light source for an extended time period. The thiol-ene based PDLC exhibited a significantly greater color change than P7 upon exposure of each PDLC to a Rayonet Reactor equipped with 350 nm light bulbs for ~ 82 hours. These initial results indicate resistance of the maleimide/vinyl ether based systems to changes upon long term light exposure.

4. Conslusions

This report clearly shows that PDLCs with excellent electro-optic switching properties can be made using initiatorless systems based upon bismalelimdes and divinyl ethers. This new type of resin system, which has been suggested for its potential use in conventional photocuring applications of clear coats, shows promise for use as the network binder in PDLCs. Fast switching times and high Contrasts were generated for PDLCs prepared from bismaleimide/divinyl ether systems where the concentration of a divinyl ether oligomer with urethane groups was relatively high. In the PDLCs where the urethane content is high, SEM micropraphs indicate formation of larger liquid crystalline droplets that span a significant portion of the film's thickness. Finally, these PDLCs based upon multifunctional maleimides and vinyl ethers are stable in simulated sunlight conditions.

Acknowledgement. This work was supported by AFOSR (DEPSCoR), NSF (DMR-9512506) and the MRSEC Program of the National Science Foundation under Award Number DMR 0213883.

References

[1] Doane, J.W., Vaz, N.A., Wu, B.G., Zumer, S., 1986, *Appl Phys. Lett.,* **48**, 269.

[2] West, J. L., 1990, *Liquid Crystalline Polymers*, ACS Symposium Series Number 435, Eds. R. Weiss and C. Ober, ACS, Washington, D. C., pp. 475-494.

[3] Doane, J.W.; Yang ,D.K.; Chie, L.C. ,1991, *Current Trends in Polymer Dispersed Liquid Crystals*, 175-178.

[4]   Wu, B-G., West, J.L., Doane, J.W. ,1987, *J. of Appl. Phys.*, **62**(9), 3925.

[5]   Whitehead, J., Zumer, S., Doane, J. W. ,1993, *J. Appl. Phys.* ,**73** , 1057.

[6]   Draziac, P. S. , 1986, *J. Appl. Phys.*, 60, 2142.

[7]   Jonsson, S., Sundell, P. E., Shimose, M., Owens, J., Hoyle, C. E., 1995, *Polym. Mater. Sci. Eng.*, **72**, 470.

[8]   Hoyle, C. E., Clark, S. C., Miller, C., Owens, J., Whitehead, J., Jonsson, S., Sundell, P. E., Shimose, M., Katogi, S., Morel, F., Decker, C., 1997, *Proc. RadTech Asia '97*, G2-6.

[9]   Jonsson, S., Sundell, P. E., Shimose, M., Clark, S., Miller, C., Morel, F., Decker, C., Hoyle, C. E., 1997, *Phys. Res. B Nuc. Inst. Met*, **131**, 276.

[10]   Kohli, P., Scranton, A.B., Blanchard, G.J., 1998, *Macromolecules*, **31**, 5681.

Chapter 13

Stimuli-Responsive Supramolecular Solids: Functional Porous and Inclusion Materials

Dmitriy V. Soldatov[1,2]

[1]Nikolaev Institute of Inorganic Chemistry, Siberian Branch of the Russian Academy of Sciences, Novosibirsk, 630090 Russia
[2]Steacie Institute for Molecular Sciences, National Research Council, Ottawa, Canada

Bulk solids that can host molecules of another, the guest component, form a large class of responsive materials. These materials can display a dramatic change in their structure and properties in the presence of various guests and, for each particular guest, the change may be individual. Two criteria are essential in the selection and design of such materials. First, the presence of weaker interactions that make it possible to organize the host molecules in a variety of mutually convertible regular structures (supramolecular organization). Second, high affinity to form inclusion compounds that is driven by thermodynamic reasons (clathration ability). Host materials of this type interact with external environment chemically (on a supramolecular level; the chemical composition of host usually does not change) and may be utilized for the creation of new generations of functional and smart materials.

Supramolecular science provided chemists with a new tool to manipulate matter and made it possible to create artificial systems competing with nature in their organization and complexity (1-8). *Supramolecular solids* exhibit a *bonding hierarchy* among components so that their properties are defined not only by covalent bonds within "building elements" but also by "supramolecular bonds" organizing these elements in a 3D structure. Also, supramolecular solids are bulk materials. According to various studies, the minimal size of a particle representing the properties of the corresponding bulk material is within 30-150 nm with 10^6-10^8 atoms or molecules (9-12). Therefore, macroscopic properties of these materials, such as stability and sorption propensity, should be studied along with their structure and composition.

The range of supramolecular materials embraces crystalline inclusion compounds and coordination polymers (13,14), binary and multicomponent molecular crystals (15,16), organic-inorganic nanocomposites (17-19), partially crystalline or glassy low-dimensional organic polymers (20), dendritic materials (21), materials formed by nanoparticles and self-assembled supramolecular layers (22-25), liquid-like systems such as gels (26-28), liquid crystals (29,30) and liquid clathrates (31), and various bio- and biomimetic materials (32-38). One of major domains are *"soft" supramolecular materials* - dynamic, responsive systems build upon weak interactions (39). Current research on supramolecular materials is an intrinsic part of many traditional disciplines, as well as relatively new ones, such as *inclusion chemistry* (13,40-42), *crystal engineering* (43-47) and na*nomaterials science* (22-25,48-50).

A tremendous choice of new materials supramolecular chemists became able to generate opened new opportunities for materials science but, on the other hand, revealed the inefficiency of "sieving-the-desert" search for a suitable material among such a variety. The new tendencies brought forward the concept of *functional materials* which is in concordance with a recent shift in materials science toward design in which researchers can predict new materials they would like to have rather than having to discover them (51). A number of emerging terms such as "responsive" (52-56), "smart" (57-61) and "intelligent" (62-67) name materials for their desired behavior, in contrast to terms used for centuries to name materials for their composition ("inorganic", "organic") or structure ("polymeric", "crystalline").

This review outlines one class of supramolecular materials - porous and inclusion solids - that have a very high potential for the utilization as functional elements in specific, practically useful devices and processing schemes. These materials may show various types of inclusion behavior. *Sorption* (68-72), a reversible inclusion of guests into a stable porous matrix, is displayed by such materials as zeolites and mesoporous solids. *Clathration* (73-76) is characteristic of clathrates, inclusion solids with unstable host framework. *Encapsulation* (trapping) (74,77-79) describes irreversible inclusion and may be observed in clathrates. *Dissolution* (80-82) accompanies sorption in the course of inclusion in some glassy and semi-crystalline organic polymers. Finally, *dynamic sorption*, a special behavior that resembles both sorption and clathration, is observed in a new family of materials known as dynamic sorbents.

Porous Solids

Porous solids possess an open system of pores that can be reversibly filled with another, the guest component in the course of sorption (68-72). According to IUPAC classification, porous solids are divided into *microporous* (< 2 nm), *mesoporous* (2–50 nm), and *macroporous* (> 50 nm) materials (68,69).

Porous structures exist whether empty or filled by guest molecules; they do not change significantly upon guest inclusion. Porous materials may communicate with external environment in a passive way, for example, selectively absorbing only certain species from the surrounding atmosphere. Microporous solids, such as *zeolites*, are ideal for this purpose demonstrating the highest energy and selectivity of sorption due to a geometrically specific cumulative effect of many interatomic contacts between the guest molecule and pore walls.

Zeolites and Related Materials

Zeolites are the oldest known family of microporous materials. Zeolites are crystalline inorganic polymers (aluminosilicates) build by covalently linked AlO_4 and SiO_4 polyhedra. Their microporous frameworks are very stable (kinetically) and therefore zeolite materials display typical sorption behavior.

General formula of zeolites is $M_n[(AlO_2)_x(SiO_2)_y]*mH_2O$. The framework is neutral for pure silica materials and negatively charged for alumina-containing structures. The metal cations M and guest water molecules reside in the cavity space of the framework. The metal cations may be replaced and the water molecules may be completely removed from the structure at elevated temperatures. The empty micropore space may be filled with various organic and inorganic species including gases, petrochemicals, metal atoms and clusters. Zeolites differ in the capacity and geometry of their micropore space and reveal a remarkable selectivity toward a guest species that provide the best fit upon inclusion.

Presently, about 50 natural zeolites are known (83) and hundreds of synthetic zeolites and "zeotypes" (structures created by replacing Al and Si with P, Ge, Fe and so on) have been synthesized (84-86). The greatest use of natural zeolites (> 1 mln tonnes) has been in radioactive waste encapsulation (mainly for ion-exchange of ^{137}Cs) after the Chernobyl disaster in 1986 (84). Synthetic zeolites are extensively used in the petroleum industry. For example, between 30 and 50% of all motor fuels (gasoline, jet, diesel) have been produced world wide with Y zeolite catalysts (85).

Successful applications of zeolitic materials stimulated the creation of a great number of microporous inorganic sorbents (87). Most applications of these materials exploit their efficiency in sorption, ion exchange and catalysis.

Organic Zeolites and Biozeolites

Organic zeolites are defined as organically based materials with permanent porosity (20). Currently, most organic zeolites fall into one of three groups: glassy or semi-crystalline organic polymers (see the following subsection), crystalline organic materials, crystalline metal-organic frameworks (MOFs).

The evident advantages of organically-based zeolite analogs is their potential diversity resulted from unlimited number of building units that can be utilized, the ease of control over self-assembly of these units in the porous structure, and the ability of the porous structure to respond individually to each particular guest.

Organic zeolites display sorption behavior mimicking true zeolites. At the same time, they reveal distinctions. First, these sorbents are essentially hydrophobic. The inner surface of the micropores is formed either by organic molecules building the solid phase or by organic ligands coordinating metal centers. Second, organic zeolites are mostly soft materials and may reveal a remarkable flexibility not seen in true zeolites. The sorption behavior of organic zeolites may be accompanied with dissolution or may resemble clathration (see section Dynamic Sorbents). In other words, the structure of organic zeolites is dynamic and may display various changes defined by the characteristics and concentration of guest species arising in the surroundings.

The concept of "organic zeolites" was inspired by "zeolitic" sorption behavior of some metal complexes examined by Barrer and coworkers (74,88,89). The term itself arose as jargon to define solids able to reversibly and selectively absorb large amounts of organic species while showing poor tendency toward sorption of inorganic guests (90). The term passed from the verbal to the literature only in 1980's and 1990's (91,92), and the first review entitled "Organic Zeolites" appeared as a short sub-section in a book in 1996 (93).

Crystalline organic zeolites created from purely organic molecules represent either van der Waals packing structures (94,95) or hydrogen-bonded frameworks (91,96-99). The pore space is organized in channels that may be either isolated or intersecting to form 2D or 3D systems; the geometry of the channels varies.

Porous MOFs present even a greater diversity of structures reviewed in (72). In addition to van der Waals interactions and hydrogen bonds, secondary and strong coordination bonds are extensively utilized. It is in this field the attempts to develop rational approaches seem to be the most successful and promising.

Biozeolites is a new family of organic zeolites made up of peptides, naturally produced and biologically compatible materials. Biozeolites are nontoxic and might be utilized in biological and medical contexts, such as in chiral recognition/separation or preservation/storage of drugs.

Studies on two bulk dipeptide materials, L-alanyl-L-valyl and L-valyl-L-alanine, showed that these crystalline solids are microporous with ~11% fraction of porous space (100). The preferential sorption even toward chemically inert species such as Xe was observed. Remarkably, the microporous forms of these dipeptides are thermodynamically stable in the absence of guest and they never collapse to a dense form. Their crystal structure reveals a hydrogen-bonding helical assembly of dipeptide molecules building channels that are essentially chiral. Görbitz demonstrated the existence of the whole family of crystalline dipeptides with hydrophobic (101) and hydrophilic (102) channel structures but the sorption ability of these materials has not been tested.

Microporous Organic Polymers

A number of 1D organic polymers exhibit complex inclusion behavior that appears to be a sum of two processes, sorption and dissolution. In addition to other qualities, these materials are convenient for processing, in particular, for the production of microporous films and membranes. The porosity of the bulk phases of these polymers arises from their special molecular geometries (Figure 1). The polymeric molecules have bulky groups attached to the main chain that restrict the conformational freedom of the molecule and prevent close packing of the molecules in a 3D structure.

Poly(trimethylsilylpropyne) **1** (Figure 1) forms glassy materials with outstanding gas permeability (103). Poly(vinyltrimethylsilane) **2**, another highly permeable glassy polymer, has found an industrial application in membrane gas separation processes (104). Syndiotactic polystyrene **3** and its variations **4** and **5** have recently become a subject of extensive studies (81,105-113). These polymers form crystalline materials with high clathration and sorption ability. Inclusion and microporous modifications of the polymers reveal helical chains running parallel and leaving significant amount of cavity space. Polymeric sulfonic acids **6** (114) and poly(phenyleneoxide) polymers **7** and **8** (80) represent other examples of sorbent materials. It should be noted that the molecular structures of the above sorbents were not deliberately designed. An example of purposefully created porous polymers is given by so-called "polymers of intrinsic porosity" (PIMs), complex structures with highly rigid and contorted molecular architectures (115).

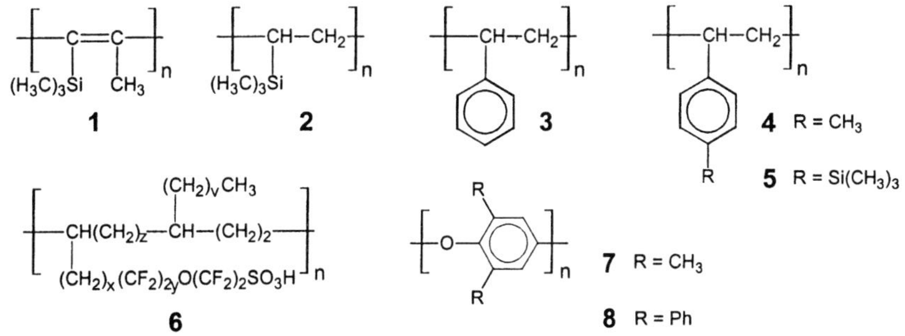

Figure 1. Molecular structures of microporous organic polymers.

The robustness of microporous structure may be enhanced by a significant restriction of the molecular conformational freedom in 2D and 3D architectures. The synthesis of dendritic structures in so-called organic aerogels (116), cross-linking of 1D polymers (117) and polymerization of synthetic lipids (118) illustrate this useful strategy.

Clathrates

Clathrates were first materials for which the realization of supramolecular organization had been extensively recognized (40-42,73,119). Currently, the term "clathrate" is synonymous with "inclusion compound" or "host-guest compound" (120). At the same time, the term "clathration" is used mostly to describe an inclusion reaction accompanied with the host structure reorganization. By this definition, host materials that form clathrates reveal the most dramatic changes of their properties in the processes of inclusion/dissociation. The half-century studies on clathrates have been comprehensively reviewed (7,13,14,40,42,121).

The empty clathrate framework is unstable and the host exists in another, dense modification in the absence of guests. The inclusion of guest may occur irreversibly, as encapsulation (74,77-79). Encapsulating materials may be utilized for the entrapment, storage and controlled release of gaseous or condensed guest species. However, in most cases the inclusion and escape of guest is accompanied with reversible transformations of the host material from dense to inclusion form and vise versa.

Some hosts, such as urea (122), cholic acid (123) tetraarylporphyrins (124,125), tetrapyridine Werner complexes (126-128) and polymeric cyanometallates (129) persistently form only one, or a very limited number, of basic inclusion architectures. The preference of such hosts toward the particular guest geometry is predictable and may be well utilized in separation technologies. For example, the narrow cylindrical channel of urea is ideal for the inclusion of unbranched saturated organics. Recent studies suggest the use of urea for fractionation of free fatty acid mixtures from seed oil as the inclusion solid phase shows preference for unsaturated and monosaturated components of the oil (130,131).

In contrast, other hosts display a variety of structures and some of them build up a distinct host framework for nearly each new guest. Gossypol, a natural product derived from the cottonseed, is an example of such hosts. Gossypol forms clathrates with all organic solvents ever tested; 44 crystal structure modifications of these clathrates present 14 different modes of hydrogen-bonding association of the host molecules in a 3D structure (132).

Finally, for some hosts the same host-guest combination may result in two to five structurally different solids, with their formation defined by thermodynamic or synthetic conditions (89,133-142).

The variability of host materials is well illustrated by metal complexes where an extensive and easy modification of the host molecule is possible without losing their ability to form clathrates. About 10 different metals (M), 15 anionic groups (X), and >100 organic ligands (A) were incorporated successfully into Werner complex host of general formula $[MA_4X_2]$ to give hundreds of new host materials, each able to include a variety of guest components (143,144). Detailed studies made it evident that for every guest, a characteristic host receptor specifically selective toward the particular guest, could be found.

The influence of guest inclusion on the host material properties may be caused by a reversible conformational or chemical change in the host species. Minor changes of the host molecular conformation may result in a drastic color change observed upon clathration (145-151). In one of those cases (148) two clathrates of a copper(II) host complex with the same guest have the host:guest molar ratios 1:2 and 1:2/3, and blue and green color, respectively. The transition temperature, ~29°C under the layer of dry guest solvent, substantially decreases in the presence of very small quantities of water because it may be included as an additional guest in the crystal structure of the green clathrate. Cis-trans isomerization of host metal complexes may occur upon inclusion of inorganic ions (152), polar molecules like water (153), or typical organic molecules such as benzene derivatives (154,155). Guest-induced oligo-/polymerization of host may trigger a dramatic change in bulk properties of the host material. Nickel(II) dibenzoylmethanate monomer, a dark-brown diamagnetic solid, reacts with benzene to give an inclusion compound, grass-green paramagnetic solid (156).

The observed changes in color and magnetic properties are caused by trimerization of the host monomer favored by the ability of the trimer to form an inclusion compound with benzene. Exposure of a polymeric copper(I) complex to toluene vapor results in the disappearance of its blue emission and the appearance of yellow emission due to the conversion of the polymeric host to tetrameric species which form an inclusion compound with guest toluene (157).

In many cases the presence of functionality in a material implies change. The ability to change is a fundamental quality of responsive and smart materials. Host solids that form clathrates present a wide and diverse class of materials where change may be easily initiated, controlled and manipulated.

Dynamic Sorbents

Dynamic sorbents is a recently developed family of porous materials that undergo a significant structural change during sorption (Figure 2). Dynamic sorption bears a certain resemblance to clathration and may be recognized by monitoring structural changes of the sorbent material. In the literature the host frameworks that allow structural deformations were referred to as dynamic (158), soft (159), or flexible (160).

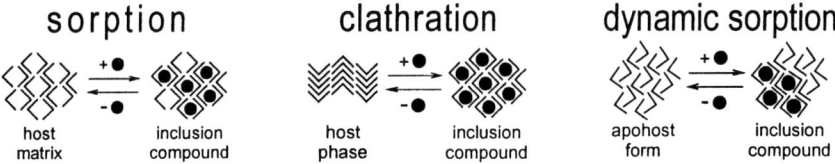

Figure 2. Sorption, clathration and dynamic sorption.

Flexible porosity is facilitated by weaker interactions in the crystal framework and requires finding an appropriate balance of strong and weak interactions within a designed architecture. The microporous form of [Ni(4-methylpyridine)$_4$(NCS)$_2$] complex is one of very few robust architectures build solely upon van der Waals interactions (89). Its collapse into a dense form requires extreme translational and rotational movement of molecules and does not occur at ambient conditions. The structure is highly flexible and "shrinks" upon guest removal, with the unit cell volume changing up to ~14%. Flexible architectures of higher dimensionality are more common. A 2.5% unit cell volume contraction reported for a 2D coordination polymer occurs due to a scissor-like shift in interlocked layers (161). A 3% contraction reported for a 3D coordination polymer is possible due to distortion in a coordination polyhedron of Ag(I) (162). More flexible 3D architectures utilized angular flexibility of secondary coordination bonds, to give 8.1% (163) and 8.6% (164) changes in the unit cell volume. An example of a reverse behavior has been also reported, where the host framework showed expansion upon removal of the guest species (165). In this case the flexibility is due to angular distortions occurring in the base of ZnO_4N coordination square pyramid.

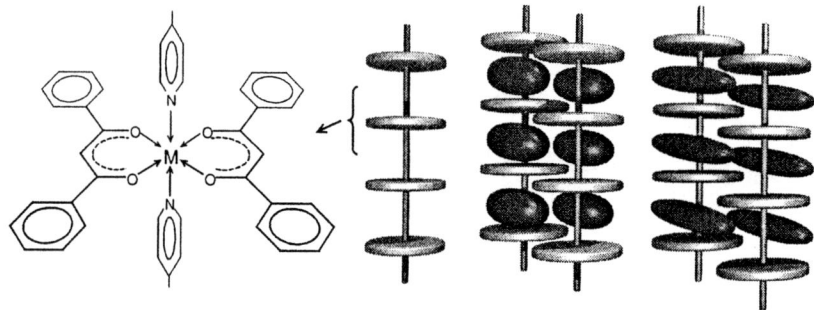

Figure 3. Molecular structure of polymeric [M(bipy)(DBM)$_2$] and schematic representation of packing patterns in inclusion compounds of the host.

One of recently designed dynamic sorbents (Figure 3) has a "ladder-and-platform" molecular geometry (166,167). The platform is constructed from a metal(II) cation chelated by two dibenzoylmethanate (DBM) anions. The whole unit is flat and relatively rigid. The platforms are connected by 4,4'-bipyridyl (bipy) bridging ligands to form a linear 1D polymer. In terms of packing, this molecular architecture is frustrating. The molecules interdigitate leaving a significant amount of cavity space available for guest species. Due to van der Waals mode of packing, the structure is very flexible. The nickel complex forms ten different crystal structures with 13 tested guest solvents (166). The host material shows either greening or yellowing upon inclusion of the different guests due to distortions in the coordination environment of Ni(II). The sorption of methane and xenon also is accompanied with noticeable structural deformations of the host crystal structure.

The actual porosity of guest-free dynamic sorbents may differ substantially from porosity estimated in sorption experiments or the crystal structure of the inclusion material. Surprisingly, some crystalline materials that look as microporous in a sorption experiment are in fact non-porous. The crystal structure of a 1D coordination polymer reported by Takamizawa and coworkers (168) reveals a dense structure without solvent molecules or porous space. Nevertheless, the bulk polymer readily absorbs gases, with the sorption isotherm being characteristic of microporous materials. The micropore volume in the material estimated from the nitrogen isotherm is 0.16 cm^3/g. Presumably, there is a structural change that occurs at an extremely low relative pressure of ~5x10^{-6}. Other studies (145,169-172) confirm structural transformations at a low "gate" pressure of guest. The empty modification of dynamic materials is referred to as *apohost* (145,169), a preorganized form instantly acquiring or increasing its porosity in response to the guest appearance.

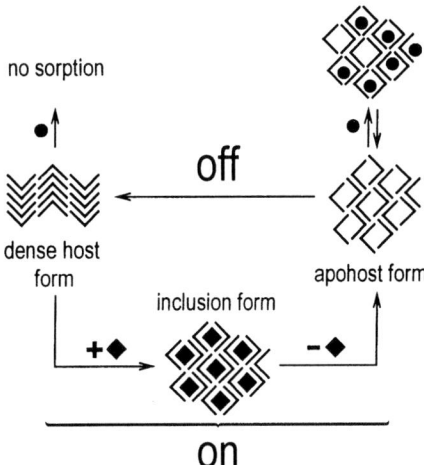

Figure 4. Processing scheme for a smart sorbent. Diamonds and circles indicate templating and working guests, respectively.

The ability of inclusion structures to be preserved in the apohost form rather than to collapse was utilized for the development of *smart sorbents* (Figure 4) (145,160,173). Microporosity in such materials may be switched on and off by an operator through the application of certain sequences of external stimuli. The operation "on" comprises two steps, the template-induced conversion of the dense host modification into an inclusion form and the removal of the guest template in flow of a neutral gas. After this operation, the host in its apohost form acquires sorption ability and may be used as a conventional sorbent. The operation "off" is accomplished simply by triggering the collapse of the apohost into the dense form using a heat pulse or a suitable catalyst.

Stepwise reversible changes of bulk physical properties of porous solids caused by the guest inclusion present another example of prospective switchable materials. A spin crossover material reported by Halder and coworkers (174) shows the temperature-induced switching that arises from the presence of Fe(II) spin crossover centers within the framework lattice. The switching temperature depends on the presence and the nature of guests included in the lattice. In another material, reported by Maspoch and coworkers (175), a step-wise, temperature induced long-range magnetic ordering possible due to organic free radicals incorporated into the host lattice. The magnetic response of this material substantially increases (up to one order of magnitude) after filling with guest.

Dynamic sorbents combine advantages of sorbents and clathrates. Similar to sorbents, they demonstrate instant sorption which does not require a significant "gate" pressure of guest. Similar to clathrates, they show significant changes in their properties upon sorption and these changes are individual for each particular guest. These solids communicate with surroundings by means of a reversible chemical process and should be taken into consideration as a special type of responsive materials.

References

1. Lehn, J.-M. *Proc. Natl. Acad. Sci. USA* **2002**, *99*, 4763-4768.
2. Pedersen, C.J.; Lehn, J.M.; Cram, D.J. *J. Inclusion Phenom.* **1988**, *6*, 337-413.
3. *Comprehensive Supramolecular Chemistry;* Atwood, J.L.; Davies, J.E.D.; MacNicol, D.D.; Vögtle, F., Eds.; Elsevier Science: New York, NY, 1996; Vols. 1-11.
4. *Encyclopedia of Supramolecular Chemistry;* Atwood, J.L.; Steed, J.W., Eds.; Marcel Dekker: New York, NY, 2004.
5. Vögtle, F. *Supramolecular Chemistry: An Introduction;* Wiley: New York, NY, 1991.
6. Lehn, J.M. *Supramolecular Chemistry: Concepts and Perspectives. A Personal Account;* VCH: New York, NY, 1995.
7. Steed, J.W.; Atwood, J.L. *Supramolecular Chemistry: An Introduction (textbook)*; Wiley: New York, NY, 2000.
8. Dodziuk, H. *Introduction to Supramolecular Chemistry;* Kluwer: Dordrecht, The Netherlands, 2002.
9. Gusev, A.I. *Phys. Usp.* **1998**, *41*, 49-76.
10. Jiang, Q.; Shi, H.X.; Zhao, M. *J. Chem. Phys.* **1999**, *111*, 2176-2180.
11. Vesnin, Y.I. *Chemistry for Sustainable Development* **2000**, *8*, 179-183.
12. Zhao, M.; Li, J.C.; Jiang, Q. *Journal of Alloys and Compounds* **2003**, *361*, 160-164.
13. *Inclusion Compounds;* Atwood, J.L.; Davies, J.E.D.; MacNicol, D.D., Eds.; Academic Press: London, 1984; Vols. 1-3; Oxford University Press: Oxford, 1991; Vols. 4-5.
14. Ref. 3, Vols. 6,7 and partly 1-3.
15. Kitaigorodsky, A.I. Molecular Crystals; Moscow: Nauka, Russia, 1971; Chapter 1, Section 20, pp 150-159 (in Russian).

16. Kitaigorodsky, A.I. Mixed Crystals; Moscow: Nauka, Russia, 1983; Chapter 11, pp 191-229 (in Russian).
17. *Special Issue on Organic-Inorganic Nanocomposite Materials*; Eckert, H.; Ward, M., Eds.; *Chem. Mater.* **2001**, *13*, Issue 10.
18. Soler-Illia, G.J.A.A.; Sanchez, C.; Lebeau, B.; Patarin, J. *Chem. Rev.* **2002**, *102*, 4093-4138.
19. Ruiz-Hitzki, E. *Chemical Record* **2003**, *3*, 88-100.
20. Soldatov, D.V.; Ripmeester, J.A. *Stud. Surf. Sci. Catal.*, in press.
21. *Dendrimers II: Architecture, Nanostructure and Supramolecular Chemistry;* Vögtle, F., Houk, K.N.; Kessler, H.; Lehn, J.-M., Eds.; *Top. Curr. Chem.*, Vol. 210; Springer Verlag, New York, NY, 2000.
22. Wang, Z.L. *Adv. Mater.* **1998**, *10*, 13-30.
23. Mann, S.; Davis, S.A.; Hall, S.R.; Li, M.; Rhodes, K.H.; Shenton, W.; Vaucher, S.; Zhang, B. *J. Chem. Soc., Dalton Trans.* **2000**, 3753-3763.
24. Shimomura, M.; Sawadaishi, T. *Curr. Opin. Colloid Interface Sci.* **2001**, *6*, 11-16.
25. Depero, L.E.; Curri, M.L. *Curr. Opin. Solid State Mater. Sci.* **2004**, *8*, 103-109.
26. Terech, P.; Weiss, R.G. *Chem. Rev.* **1997**, *97*, 3133-3159.
27. Esch, J.H.; Feringa, B.L. *Angew. Chem., Int. Ed.* **2000**, *39*, 2263-2266.
28. Luboradzki, R.; Pakulski, Z. *Tetrahedron* **2004**, *60*, 4613-4616.
29. *Handbook of Liquid Crystals;* Demus, D.; Goodby, J.W.; Gray, G.W.; Spiess, H.-W.; Vill, V., Eds.; Wiley-VCH: Weinheim, Germany, 1998; Vols. 1-3.
30. Brunsveld, L.; Folmer, B.J.B.; Meijer, E.W.; Sijbesma, R.P. *Chem. Rev.* **2001**, *101*, 4071-4097.
31. Atwood, J.L. Liquid clathrates. In: *Inclusion Compounds;* Atwood, J.L.; Davies, J.E.D.; MacNicol, D.D., Eds.; Academic Press: London, UK, 1984; Vol. 1, pp 375-405.
32. *Supramolecular and Colloidal Structures in Biomaterials and Biosubstrates;* Lal, M.; Lillford, P.J.; Naik, V.M.; Prakash, V., Eds.; Imperial College Press, London, UK, 2000.
33. Stewart, G.T. *Liquid Crystals* **2003**, *30*, 541-557.
34. Stewart, G.T. *Liquid Crystals* **2004**, *31*, 443-471.
35. Tu, R.S.; Tirrell, M. *Advanced Drug Delivery Reviews* **2004**, *56*, 1537-1563.
36. Da Silva, E.; Lazar, A.N.; Coleman, A.W. *Journal of Drug Delivery Science and Technology* **2004**, *14*, 3-20.
37. Ratner, B.D.; Bryant, S.J. *Annu. Rev. Biomed. Eng.* **2004**, *6*, 41-75.
38. Drotleff, S.; Lungwitz, U.; Breunig, M.; Dennis, A.; Blunk, T.; Tessmar, J.; Göpferich, A. *Eur. J. Pharm. Biopharm.* **2004**, *58*, 385-407.
39. Soldatov, D.V. *J. Inclusion Phenom.* **2004**, *48*, 3-9.

40. Davies, J.E.D.; Kemula, W.; Powell, H.M.; Smith, N.O. *J. Inclusion Phenom.* **1983**, *1*, 33-44.
41. Davies, J.E.D. *J. Inclusion Phenom.* **1998**, *32*, 499-504.
42. Dyadin, Yu.A.; Terekhova, I.S.; Rodionova, T.V.; Soldatov, D.V. *J. Struct. Chem.* **1999**, *40*, 645-653.
43. *Solid-State Supramolecular Chemistry: Crystal Engineering*; MacNicol, D.D.; Toda, F.; Bishop, R., Eds. In: *Comprehensive Supramolecular Chemistry;* Elsevier Science: New York, NY, 1996; Vol. 6.
44. Desiraju, G.R. *Crystal Engineering: the Design of Organic Solids;* Elsevier: Amsterdam, The Netherlands, 1989.
45. Blake, A.J.; Champness, N.R.; Hubberstey, P.; Li, W.-S.; Withersby, M.A.; Schröder, M. *Coord. Chem. Rev.* **1999**, *183*, 117-138.
46. Moulton, B.; Zaworotko, M.J. *Advances in Supramolecular Chemistry* **2000**, *7*, 235-283.
47. Biradha, K. *CrystEngComm* **2003**, *5*, 374-384.
48. *Nanoporous Materials III;* Sayari, A.; Jaroniec, M., Eds.; In: *Studies in Surface Science and Catalysis;* Elsevier: Amsterdam, The Netherlands, 2002, Vol. 141.
49. Sayari, A.; Liu, P. *Microporous Materials* **1997**, *12*, 149-177.
50. Sayari, A.; Hamoudi, S. *Chem. Mater.* **2001**, *13*, 3151-3168.
51. Olson, G.V. *Science* **2000**, *288*, 993-998.
52. Barisci, J.N.; Lewis, T.W.; Spinks, G.M.; Too, C.O.; Wallance, G.G. *Journal of Intelligent Materials Systems and Structures* **1998**, *9*, 723-731.
53. Russel, T.P. *Science* **2002**, *297*, 964-967.
54. Sershen, S.; West, J. *Advanced Drug Delivery Reviews* **2002**, *54*, 1225-1235.
55. Luzinov, I.; Minko, S.; Tsukruk, V.V. *Progr. Polim. Sci.* **2004**, *29*, 635-698.
56. Rajagopal, K.; Schneider, J.P. *Curr. Opin. Struct. Biol.* **2004**, *14*, 480-486.
57. Newnham, R.E. *Mater. Res. Soc. Bull.* **1997**, *22 (5)*, 20-34.
58. Newnham, R.E. *Acta Crystallogr. A* **1998**, *54*, 729-737.
59. Cao, W.; Cudney, H.H.; Waser, R. *Proc. Natl. Acad. Sci. USA* **1999**, *96*, 8330-8331.
60. Mackerle, J. *Modelling Simul. Mater. Sci. Eng.* **2003**, *11*, 707-744.
61. Soldatov, D.V. Soft and smart materials. In: *Encyclopedia of Supramolecular Chemistry;* Atwood, J.L.; Steed, J.W., Eds.; Marcel Dekker: New York, NY, 2004; pp 1302-1306.
62. Rogers, C.A. *Scientific American* **1995**, *September Issue*, 154-157.
63. Hornbogen, E.; Mertmann, M. *Metall* **1996**, *50*, 809-814.
64. Holtz, J.H.; Asher, S.A. *Nature* **1997**, *389*, 829-832.
65. Takagi, T. *Journal of Intelligent Materials Systems and Structures* **1999**, *10*, 575-581.

66. Sakiyama-Elbert, S.E.; Hubbell, J.A. *Annu. Rev. Mater. Res.* **2001**, *31*, 183-201.
67. Hilt, J.Z. *Advanced Drug Delivery Reviews* **2004**, *56*, 1533-1536.
68. Sing, K.S.W.; Everett, D.H.; Haul, R.A.W.; Moscou, L.; Pierotti, R.A.; Rouquerol, J.; Siemieniewska, T. *Pure Appl. Chem.* **1985**, *57*, 603-619.
69. Rouquerol, J.; Avnir, D.; Fairbridge, C.W.; Everett, D.H.; Haynes, J.H.; Pernicone, N.; Ramsay, J.D.F.; Sing, K.S.W.; Unger, K.K. *Pure Appl. Chem.* **1994**, *66*, 1739-1758.
70. Brunauer, S. *The Adsorption of Gases and Vapors;* Princeton University Press: Princeton, NJ, 1943; Vol. 1.
71. Kruk, M.; Jaroniec, M. *Chem. Mater.* **2001**, *13*, 3169-3183.
72. Kitagawa, S.; Kitaura, R.; Noro, S. *Angew. Chem., Int. Ed.* **2004**, *43*, 2334-2375.
73. Powell, H.M. *J. Chem. Soc.* **1948**, 61-73.
74. Barrer, R.M. Porous crystals. Clathration, trapping and zeolitic sorption. In: *Molecular Sieves;* Meier, W.M.; Uytterhoeven, J.B., Eds.; Adv. Chem. Ser.; American Chem. Soc.: Washington, NY, 1973; Vol. 121, pp 1-28.
75. Waals, J.H. van der; Platteeuw, J.C. *Adv. Chem. Phys.* **1959**, *2*, 1-57.
76. Belosludov, V.R.; Lavrent'ev, M.Yu.; Dyadin, Yu.A. *J. Inclusion Phenom.* **1991**, *10*, 399-422.
77. Atwood, J.L.; Barbour, L.J.; Jerga, A. *Science* **2002**, *296*, 2367-2369.
78. Enright, G.D.; Udachin, K.A.; Mourakovski, I.L.; Ripmeester, J.A. *J. Am. Chem. Soc.* **2003**, *125*, 9896-9897.
79. Rudkevich, D.M.; Leontiev, A.V. *Aust. J. Chem.* **2004**, *57*, 713-722.
80. Ilinitch, O.M.; Fenelonov, V.B.; Lapkin, A.A.; Okkel, L.G.; Terskikh, V.V.; Zamaraev, K.I. *Micropor. Mesopor. Mater.* **1999**, *31*, 97-110.
81. Hodge, K.; Prodpran, T.; Shenogina, N.B.; Nazarenko, S. *J. Polimer Sci. B* **2001**, *39*, 2519-2538.
82. Klopffer, M.H.; Flaconnèche, B. *Oil & Gas Science and Technology – Rev. IFP* **2001**, *56*, 223-244.
83. Armbruster, T.; Gunter, M.E. Crystal structures of natural zeolites. In: *Reviews in Mineralogy and Geology;* Bish, D.L.; Ming, D.W., Eds.; Mineralogical Society of America: Washington, NY, 2001; Vol. 45, pp 1-67.
84. Tomlinson, A.A.G. *Modern Zeolites: Structure and Function in Detergents and Petrochemicals.* In: *Materials Science Foundations;* Fisher, D.J.; Wohlbier, F.H., Eds.; Trans. Tech. Publications Ltd.: Uetikon-Zuerich, Switzerland, 1998; Vol. 3.
85. Szostak, R. *Molecular Sieves: Principles of Synthesis and Identification.* Blackie Acad. & Prof.: London, UK, 1998.
86. Francis, R.J.; O'Hare, D. *J. Chem. Soc., Dalton Trans.* **1998**, 3133-3148.
87. Cheetham, A.K.; Férey, G.; Loiseau, T. *Angew. Chem., Int. Ed.* **1999**, *38*, 3268-3292.

88. Allison, S.A.; Barrer, R.M. *J. Chem. Soc. A* **1969**, 1717-1723.
89. Soldatov, D.V.; Enright, G.D.; Ripmeester, J.A. *Cryst. Growth. Des.* **2004**, *4*, 1185-1194.
90. Lipkowski, J. Physicochemical characteristic of organic zeolite behavior. In: *Proceedings of the XIth International Symposium on Supramolecular Chemistry*, Fukuoka, Japan, 2000; p 64.
91. Ibragimov, B.T.; Talipov, S.A. *J. Inclusion Phenom.* **1994**, *17*, 317-324.
92. Lipkowski, J. Werner clathrates: guest - lattice intermolecular interactions. In: *Organic Crystal Chemistry;* Garbarczyk, J.B.; Jones, D.W., Eds.; Oxford University Press: Oxford, UK, 1991; pp 27-35.
93. Lipkowsi, J. Werner Clathrates. In: *Comprehensive Supramolecular Chemistry;* MacNicol, D.D.; Toda, F.; Bishop, R., Eds.; Elsevier Science: New York, NY, 1996; Vol. 6, pp 691-714.
94. Sozzani, P.; Comotti, A.; Simonutti, R.; Meersmann, T.; Logan, J.W.; Pines, A. *Angew. Chem., Int. Ed.* **2000**, *39*, 2695-2699.
95. Plattner, D.A.; Beck, A.K.; Neuburger, M. *Helv. Chim. Acta* **2002**, *85*, 4000-4011.
96. Venkataraman, D.; Lee, S.; Zhang, J.; Moore, J.S. *Nature* **1994**, *371*, 591-593.
97. Ung, A.T.; Gizachew, D.; Bishop, R.; Scudder, M.L.; Dance, I.G.; Craig, D.C. *J. Am. Chem. Soc.* **1995**, *117*, 8745-8756.
98. Brunet, P.; Simard, M.; Wuest, J. D. *J. Am. Chem. Soc.* **1997**, *119*, 2737-2738.
99. Russel, V.A.; Evans, C.C.; Li, W.; Ward, M.D. *Science* **1997**, *276*, 575-579.
100. Soldatov, D.V.; Moudrakovski, I.L.; Ripmeester, J.A. *Angew. Chem., Int. Ed.* **2004**, *43*, 6308-6311.
101. Görbitz, C.H. *New. J. Chem.* **2003**, *27*, 1789-1793.
102. Görbitz, C.H. *Chem. Eur. J.* **2001**, *7*, 5153-5159.
103. Fried, J.R.; Goyal, D.K. *J. Polim. Sci. B* **1998**, *36*, 519-536.
104. Nametkin, N.S.; Durgaryan, S.G. *Plast. Massy* **1980**, Issue 11, 13-27 (in Russian).
105. Tsutsui, K.; Tsujita, Y.; Yoshimizu, H.; Kinoshita, T. *Polymer* **1998**, *39*, 5177-5182.
106. Milano, G.; Venditto, V.; Guerra, G.; Cavallo, L.; Ciambelli, P.; Sannino, D. *Chem. Mater.* **2001**, *13*, 1506-1511.
107. Handa, Y.P.; Zhang, Z.; Nawaby, V.; Tan, J. *Cell. Polym.* **2001**, *20*, 241-253.
108. Trezza, E.; Grassi, A. *Macromol. Rapid Commun.* **2002**, *23*, 260-263.
109. Yoshioka, A.; Tashiro, K. *Macromolecules* **2003**, *36*, 3593-3600.
110. Loffredo, F.; Pranzo, A.; Venditto, V.; Longo, P.; Guerra, G. *Macromol. Chem. Phys.* **2003**, *204*, 859-867.

111. De Rosa, C.; Rizzo, P.; Ruiz de Ballesteros, O.; Petraccone, V.; Guerra, G. *Polymer* **1999**, *40*, 2103-2110.
112. Rizzo, P.; Ruiz de Ballesteros, O.; De Rosa, C.; Auriemma, F.; La Camera, D.; Petraccone, V.; Lotz, B. *Polymer* **2000**, *41*, 3745-3749.
113. Khotimskii, V.S.; Filippova, V.G.; Bryantseva, I.S.; Bondar, V.I.; Shantarovich, V.P.; Yampolskii, Y.P. *J. Appl. Polym. Sci.* **2000**, *78*, 1612-1620.
114. Yang, Z.-Y.; Wang, L.; Drysdale, N.; Doyle, M.; Sun, Q.; Choi, S.K. *Angew. Chem., Int. Ed.* **2003**, *42*, 5462-5464.
115. Budd, P.M.; Ghanem, B.S.; Makhseed, S.; McKeown, N.B.; Msayib, K.J.; Tattershall, C.E. *Chem. Commun.* **2004**, 230-231.
116. LeMay, J.D.; Hopper, R.W.; Hrubesh, L.W.; Pekala, R.W. *Mater. Res. Soc. Bull.* **1990**, *15 (12)*, 19-45.
117. Gorbatchuk, V.V.; Mironov, N.A.; Solomonov, B.N.; Habicher, W.D. *Biomacromolecules* **2004**, *5*, 1615-1623.
118. Srisiri, W.; O'Brien, D.F.; Orädd, G.; Persson, S.; Lindblom, G. *Polym. Prepr.* **1997**, *38*, 906-907.
119. Powell, H.M.; Riesz, P. *Nature* **1948**, *161*, 52-53.
120. Dyadin, Yu.A.; Terekhova, I.S. Classical descriptions of inclusion compounds. In: *Encyclopedia of Supramolecular Chemistry;* Atwood, J.L.; Steed, J.W., Eds.; Marcel Dekker: New York, NY, 2004; pp 253-260.
121. Powell, H.M. Clathrate compounds. In: *Non-stoichiometric Compounds;* Mandelcorn, L., Ed.; Academic Press: New York, NY, 1964; pp 398-450.
122. Takemoto, K.; Sonoda, N. Inclusion compounds of urea, thiourea and selenourea. In: *Inclusion Compounds;* Atwood, J.L.; Davies, J.E.D.; MacNicol, D.D., Eds.; Academic Press: London, UK, 1984; Vol. 2, pp 44-67.
123. Nakano, K.; Sada, K.; Kurozumi, Y.; Miyata, M. *Chem. Eur. J.* **2001**, *7*, 209-220.
124. Byrn, M.P.; Curtis, C.J.; Goldberg, I.; Hsiou, Yu.; Khan, S.I.; Sawin, P.A.; Tendick, S.K.; Strouse, C.E. *J. Am. Chem. Soc.* **1991**, *113*, 6549-6557.
125. Byrn, M.P.; Curtis, C.J.; Hsiou, Y.; Khan, S.I.; Sawin, P.A.; Tendick, S.K.; Terzis, A.; Strouse, C.E. *J. Am. Chem. Soc.* **1993**, *115*, 9480-9497.
126. Soldatov, D.V.; Lipkowski, J. *J. Struct. Chem.* **1995**, *36*, 979-982.
127. Soldatov, D.V.; Ripmeester, J.A. *Supramol. Chem.* **1998**, *9*, 175-181.
128. Soldatov, D.V.; Enright, G. D.; Ripmeester, J. A.; Lipkowski, J.; Ukraintseva, E. A. *J. Supramol. Chem.* **2001**, *1*, 245-251.
129. Iwamoto, T. *J. Inclusion Phenom.* **1996**, *24*, 61-132.
130. Hayes, D.G.; Alstine, J.V.; Satterwall, F. *J. Am. Oil Chem. Soc.* **2000**, *77*, 207-213.
131. Hayes, D.G.; Alstine, J.V.; Asplund, A.-L. *Sep. Sci. Technol.* **2001**, *36*, 45-58.

132. Ibragimov, B.T.; Talipov, S.A. *Zh. Strukt. Khim.* **1999**, *40*, 849-871.
133. Lipkowski, J.; Soldatov, D.V.; Kislykh, N.V.; Pervukhina, N.V.; Dyadin, Yu.A. *J. Inclusion Phenom.* **1994**, *17*, 305-316.
134. Lipkowski, J.; Soldatov, D.V. *J. Inclusion Phenom.* **1994**, *18*, 317-329.
135. Nakano, K.; Sada, K.; Miyata, M. *Chem. Commun.* **1996**, 989-990.
136. Beketov, K.M.; Ibragimov, B.T.; Talipov, S.A. *J. Inclusion Phenom.* **1997**, *28*, 141-150.
137. Ibragimov, B.T. *J. Inclusion Phenom.* **1999**, *34*, 345-353.
138. Ibragimov, B.T.; Makhkamov, K.K.; Beketov, K.M. *J. Inclusion Phenom.* **1999**, *35*, 583-593.
139. Paternostre, L.; Damman, P.; Dosière, M. *Macromolecules* **1999**, *32*, 153-161.
140. Aladko, L.S.; Dyadin, Yu.A.; Mikina, T.V. *Russ. J. Gen. Chem.* **2002**, *72*, 364-367.
141. Aladko, L.S.; Dyadin, Yu.A.; Mikina, T.V. *Russ. J. Gen. Chem.* **2003**, *73*, 503-506.
142. Sidhu, P.S.; Enright, G.D.; Udachin, K.A.; Ripmeester, J.A. *Cryst. Growth. Des.* **2004**, *4*, 1249-1257.
143. Lipkowski, J. Inclusion compounds formed by Werner MX_2A_4 coordination complexes. In: *Inclusion Compounds;* Atwood, J.L.; Davies, J.E.D.; MacNicol, D.D., Eds.; Academic Press: London, UK, 1984; Vol. 1, pp 59-103.
144. Hanotier, J.; Radzitzki, P. Inclusion compounds of diisothiocyanato-tetrakis(α-arylalkylamine) nickel(II) complexes. In: *Inclusion Compounds;* Atwood, J.L.; Davies, J.E.D.; MacNicol, D.D., Eds.; Academic Press: London, UK, 1984; Vol. 1, pp 104-134.
145. Soldatov, D.V.; Ripmeester, J.A. *Chem. Eur. J.* **2001**, *7*, 2979-2994.
146. Soldatov, D.V.; Grachev, E.V.; Lipkowski, J. *J. Struct. Chem.* **1996**, *37*, 658-665.
147. Kaftory, M.; Taycher, H.; Botoshansky, M. *J. Chem. Soc., Perkin Trans. 2*, **1998**, 407-412.
148. Lipkowski, J.; Kislykh, N.V.; Dyadin, Yu.A.; Sheludyakova, L.A. *J. Struct. Chem.* **1999**, *40*, 772-780.
149. Soldatov, D.V.; Suwinska, K.; Lipkowski, J.; Ogienko, A.G. *J. Struct. Chem.* **1999**, *40*, 781-789.
150. Scott, J.; Asami, M.; Tanaka, K. *New J. Chem.* **2002**, *26*, 1822-1826.
151. Soldatov, D.V.; Enright, G.D.; Ripmeester, J.A. *Chem. Mater.* **2002**, *14*, 348-356.
152. Guilard, R.; Siri, O.; Tabard, A.; Broeker, G.; Richard, P.; Nurco, D.J.; Smith, K.M. *J. Chem. Soc., Dalton Trans.* **1997**, 3459-3463.
153. Adams, R.P.; Allen, H.C.; Rychlewska, U.; Hodgson, D.J. *Inorg. Chim. Acta* **1986**, *119*, 67-74.

154. Nassimbeni, L.R.; Niven, M.L.; Zemke, K.J. *Acta Crystallogr.* **1986**, *B42*, 453-461.
155. Soldatov, D.V.; Ripmeester, J.A.; Shergina, S.I.; Sokolov, I.E.; Zanina, A.S.; Gromilov, S.A.; Dyadin, Yu.A. *J. Am. Chem. Soc.* **1999**, *121*, 4179-4188.
156. Soldatov, D.V.; Henegouwen, A.T.; Enright, G.D.; Ratcliffe, C.I.; Ripmeester, J.A. *Inorg. Chem.* **2001**, *40*, 1626-1636.
157. Cariati, E.; Bu, X.; Ford, P.C. *Chem. Mater.* **2000**, *12*, 3385-3391.
158. Kitagawa, S.; Kondo, M. *Bull. Chem. Soc. Jpn.* **1998**, *71*, 1739-1753.
159. Holman, K.T.; Pivovar, A.M.; Swift, J.A.; Ward, M.D. *Acc. Chem. Res.* **2001**, *34*, 107-118.
160. Soldatov, D.V.; Ripmeester, J.A. *Stud. Surf. Sci. Catal.* **2002**, 141, 353-362.
161. Kepert, C.J.; Rosseinsky, M.J. *Chem. Commun.* **1999**, 375-376.
162. Abrahams, B.F.; Jackson, P.A.; Robson, R. *Angew. Chem., Int. Ed.* **1998**, *37*, 2656-2659.
163. Lu, J.Y.; Babb, A.M. *Chem. Commun.* **2002**, 1340-1341.
164. Soldatov, D.V.; Grachev, E.V.; Ripmeester, J.A. *Cryst. Growth. Des.* **2002**, *2*, 401-408.
165. Dybtsev, D.N.; Chun, H.; Kim, K. *Angew. Chem., Int. Ed.* **2004**, *43*, 5033-5036.
166. Soldatov, D.V.; Moudrakovski, I.L.; Ratcliffe, C.I.; Dutrisac, R.; Ripmeester, J.A. *Chem. Mater.* **2003**, *15*, 4810-4818.
167. Soldatov, D.V.; Ripmeester, J.A. *Mendeleev Commun.* **2004**, 101-103.
168. Takamizawa, S.; Hiroki, T.; Nakata, E.; Mochizuki, K.; Mori, W. *Chem. Lett.* **2002**, 1208-1209.
169. Aoyama, Y. *Top. Curr. Chem.* **1998**, *198*, 132-161.
170. Li, D.; Kaneko, K. *Chem. Phys. Lett.* **2001**, *335*, 50-56.
171. Seki, K.; Mori, W. *J. Phys. Chem. B* **2002**, *106*, 1380-1385.
172. Takamizawa, S.; Nakata, E.; Yokoyama, H.; Mochizuki, K.; Mori, W. *Angew. Chem., Int. Ed.* **2003**, *42*, 4331-4334.
173. Nossov, A.V.; Soldatov, D.V.; Ripmeester, J.A. *J. Am. Chem. Soc.* **2001**, *123*, 3563-3568.
174. Halder, G.J.; Kepert, C.J.; Moubaraki, B.; Murray, K.S.; Cashion, J.D. *Science* **2002**, *298*, 1762-1765.
175. Maspoch, D.; Ruiz-Molina, D.; Wurst, K.; Domingo, N.; Cavallini, M.; Biscarini, F.; Tejada, J.; Rovira, C.; Veciana, J. *Nature Mater.* **2003**, *2*, 190-195.

Indexes

Author Index

Akgun, Bulent, 55
Arda, Ertan, 137
Bombalski, Lindsay, 28
Boyes, Stephen G., 55
Brittain, William J., 55
Caplan, Adam, 55
Convertine, Anthony J., 43
Cusick, Brian, 28
Cyrus, Crystal, 55
Dreher, W. Reid, 122
Hladik, Molly L., 195
Houbenov, Nikolay, 68
Hoyle, Charles E., 195
Huang, Jinyu, 28
Ionov, Leonid, 68
Jérôme, C., 84
Jérôme, R., 84
Klep, Viktor, 68
Kowalewski, Tomasz, 28
Lestage, David J., 122
Lim, G. T., 166
Liu, Roger C. W., 107
Lowe, Andrew B., 43
Luzinov, Igor, 68

Matyjaszewski, Krzysztof, 28
McCormick, Charles L., 43
Minko, Sergiy, 68
Mirous, Brian, 55
Nahmad-Achar, Eduardo, 181
Pekcan, Önder, 137
Pietrasik, Joanna, 28
Pyun, Jeffrey, 28
Reddy, J. N., 166
Scales, Charles W., 43
Senyurt, Askym F., 195
Soldatov, Dmitriy V., 214
Stamm, Manfred, 68
Sue, H.-J., 166
Sumerlin, Brent S., 43
Sydorenko, Alexander, 68
Urban, Anna M., 1
Urban, Marek W., 1, 122
Vitela, Javier E., 181
Warren, Garfield T., 195
Werneth, Charles M., 195
Whitehead, Joe B., 195
Winnik, Françoise M., 107
Zdyrko, Bogdan, 68

Subject Index

A

ABAQUS®
  finite element analysis, 167–168, 178
  See also Finite element modeling for scratch damage
Acrylate containing polymers, electrografting, 98
Acrylate functional monomers, electrografting, 90–92
Adaptive re-meshing, scratch modeling, 173
Adhesion
  key feature of electrografting, 85
  origin, 88
  surface-interfacial phenomenon, 13–14
Air/water interface. *See* Poly(*N*-isopropylacrylamide) (PNIPAM)
Algorithm, scratch modeling, 172
Alkyd enamels
  artificial neural networks, 188–189
  predicting power, 190*t*
  sensitivity analysis, 190–191, 192*t*
Anchoring groups, electrografting, 95, 97
Architecture control. *See* Stimuli responsive polymer brushes
Artificial neural networks (ANN's)
  breakdown and distribution of enamel formulations, 188*t*
  determination of n, 189
  efficiency calculation, 187
  expert systems, 183
  feed forward ANN's, 185
  fundamentals, 183–187
  future work, 193
  methodology and application, 188–189

motivation, 182–183
network for prediction, 187*f*
over-training, 186–187
predicting power by enamel family, 190*t*
prediction, 187
probabilistic methods, 193
schematic of response of neuron in network, 184*f*
sensitivity analysis, 190–191
sensitivity analysis for gloss, 192*t*
simple multi-layered network, 185*f*
training, 185–186
Atomic force microscopy (AFM)
  polyelectrolyte brushes, 65, 66*f*
  poly(2-hydroxyethyl methacrylate) (pHEMA) grafts, 37, 39*f*, 40*f*
  polystyrene grafted layers at different grafting densities, 74*f*
  Si–OH vs. Si–H substrates with poly(*N,N*-dimethylaminoethyl methacrylate) (pDMAEMA) layer, 35, 36*f*
  surface-interfacial phenomena, 10–11
  tapping mode AFM of poly(dimethylsiloxane)-*b*-poly(styrene)-*b*-poly[(1-dimethoxymethylsilyl)propyl acrylate] (pDMS-*b*-pSt-*b*-pDMSA) brushes, 31*f*
Atom transfer radical polymerization (ATRP)
  brushes from oxidized silicon wafers, 33–34
  copolymer synthesis of poly(*N,N*-dimethylaminoethyl methacrylate)-*b*-poly(trimethoxysilylpropyl

methacrylate) [pDMAEMA-*b*-pTMSPMA], 32
polyelectrolyte diblock copolymer brush, 58–60
polyelectrolyte homopolymer brush, 57–58
sequential electrografting and, 94
surface-initiated ATRP of pDMAEMA from silicon wafers, 34–37
synthetic technique for well-defined polymers, 29, 56
See also Stimuli responsive polymer brushes
Azobenzene, conjugated polymers, 18–19

B

Backbone activation energies, annealing time of films, 155–156
Bilayered films, electrografting, 92–95
Bio-absorption or bio-adsorption, recognition and response driven systems, 2
Biocompatibility, 2-hydroxyethyl methacrylate copolymers, 8
Biomaterials, stimuli-responsive polymers, 2
Biomedical materials, surface-interfacial phenomenon, 10
Bio-sensing, reversible addition fragmentation chain transfer polymers, 50, 52
Biozeolites, porous solids, 218
Bismaleimide/divinyl ether polymer dispersed liquid crystals (PDLCs)
electro-optic behavior, 206–209
See also Polymer dispersed liquid crystals (PDLCs)
Block copolymer brushes. *See* Stimuli responsive polymer brushes
Bonding hierarchy, supramolecular solids, 215
Brushes
electrografting and nitroxide-mediated polymerization (NMP), 94–95
mixed gradient, 76–81
See also Polyelectrolyte brushes; Polymer brushes

C

Cathodic electrografting
applications, 100–101
chemisorption, 89
electrografting on conducting substrates, 90
electrografting mechanism, 85, 87–88
experimental facts, 85
extension to (meth)acrylic monomers other than acrylonitrile (AN), 88–90
fundamentals, 85–90
origin of adhesion, 88
polyacrylonitrile (PAN) on common metal, 85, 86*f*
relation between (meth)acrylate monomers and solvent of lower donor number (DN), 88, 89*t*
schematic mechanism of electrografting of AN, 87
See also Electrografting; Electro-responsive coatings
Ceramics, magnetic fields, 19–20
Characterization, adhesion, 88
Chemisorption, solvents and electropolymerization, 89
Chiroptical response, azobenzene modified helical polymers, 18–19
Clathrates, supramolecular solids, 219–221
Clathration, inclusion, 215
Coalescence
knowledge about, 123
See also Colloidal dispersions

Coatings. *See* Electro-responsive coatings
Colloidal dispersions
  coalescence, 123
  crosslinking and particle morphology, 125–127
  film formation from, 123
  interactions between –SO$_3^-$Na$^+$ groups of SDS and acid groups in presence of copolymers, 130f
  interactions in presence of fluorosurfactants, 132, 133f
  interfacial interactions, 124–125
  local ionic clusters, 129, 131
  migrational responses of surfactants during film formation, 126–127
  preferential orientation of –SO$_3^-$Na$^+$ groups near copolymer surface, 126f
  schematic illustrating vibrational dependencies of sodium dioctyl sulfosuccinate (SDOSS), 124f
  schematic of film formation process, 128f
  schematic of mobility of SDOSS and water during film formation, 127f
  schematic of particle interactions, 131f
  surfactant-particle interactions, 127–129
Compatibility, 2-hydroxyethyl methacrylate copolymers, 8
Compression isotherms. *See* Poly(*N*-isopropylacrylamide) (PNIPAM)
Conducting polymers, electrografting, 91–92
Conducting substrates, electrografting on, 90
π-Conjugated systems, magnetic fields, 19
Contact algorithm, scratch modeling, 172

Contact angle, gradient grafted polymer layers, 72–74
Controlled radical polymerization. *See* Reversible addition fragmentation chain transfer (RAFT)
Copolymer
  local ionic clusters, 129–131
  synthesis of polyelectrolyte brushes, 58–60
  See also Colloidal dispersions; Reversible addition fragmentation chain transfer (RAFT)
Copolymer brush, electrografting and nitroxide-mediated polymerization, 94–95
Cracking criterion, scratch modeling, 171–172
Crazing
  scratch damage process, 176, 178
  scratch modeling, 171–172
Crossing density, model, 142, 155
Cross-linked polymers
  responses to environmental changes, 10
  uses, 5
Crosslinking, mobility of surfactants, 125–126
Crystalline phase orientation, uniaxial elongation, 3–4

D

Debonding, scratch modeling, 171–172
Dewetting, surface-interfacial phenomenon, 13–14
Diblock copolymer, synthesis of polyelectrolyte brushes, 58–60
N-[3-(Dimethylamino)propyl] methacrylamide (DMAPMA), reversible addition fragmentation chain transfer polymerization, 48, 49f

3-(N,N-Dimethylvinylbenzylammonio)propanesulfonate (DMVBAPS)
  chemical structure, 45f
  reversible addition fragmentation chain transfer (RAFT), 44
Direct energy transfer (DET)
  latex film preparations, 144–145
  studying polymer interdiffusion, 138–139
  See also Latex film formation
Dispersions. *See* Colloidal dispersions
Dissolution, inclusion, 216
Dynamic analysis, scratch step, 169
Dynamic sorbents
  actual porosity, 222–223
  advantages, 224
  flexible porosity, 221–222
  ladder-and-platform molecular geometry, 222
  microporosity, 223
  processing scheme for smart sorbent, 223f
  schematic of packing patterns, 222f
  smart sorbents, 223
  sorption, clathration and dynamic sorption, 221f
  stepwise reversible changes of physical properties, 223–224
  See also Supramolecular solids
Dynamic sorption, inclusion, 216

E

Electric field, liquid crystalline materials, 17–18
Electrografting
  AFM (atomic force microscopy) images of poly(n-butyl acrylate)-b-polystyrene chains at stainless steel surface, 96f
  anchoring groups, 95, 97
  applications, 100–101
  examples of suitable acrylate containing polymers, 98
  inimers or production of bilayered films, 92–95
  poly(ε-caprolactone) (PCL), 99
  poly(ethylene oxide) (PEO) or poly(ethyl acrylate) (PEA), 99
  reactive polymers, 97–100
  sequential, and atom transfer radical polymerization (ATRP), 94
  sequential, and nitroxide-mediated polymerization (NMP), 94–95
  sequential, and ring-opening metathesis polymerization (ROMP) of norbornene and derivatives, 93
  sequential, and ring-opening polymerization (ROP) of lactides and lactones, 93–94
  See also Cathodic electrografting; Electro-responsive coatings
Electrolyte-responsive polymers, reversible addition fragmentation chain transfer, 44
Electro-optic measurements
  bismaleimide/divinyl ether based polymer dispersed liquid crystals (PDLCs), 206–209
  diagram of set up for, of PDLC cells, 201f
  procedure, 200
  thiol-ene based PDLCs, 209
  See also Polymer dispersed liquid crystals (PDLCs)
Electro-responsive coatings
  electroactivity of poly(ethylacrylate)/poly(pyrrole) (PEA/PPy) binary film, 91–92
  electroactivity of PPy film, 91–92
  electrografting of acrylates precursor of conducting polymers, 91–92

241

electrografting of inimers or production of bilayered films, 92–95
new functional acrylates for electrografting, 90, 91
PPy nanowires, 92, 93f
See also Cathodic electrografting; Electrografting

Elongation
crystalline phase orientation, 3–4
morphological changes, 4
negative Poisson ratio (NPR), 4

Enamels
artificial neural networks, 188–189
predicting power, 190t

Encapsulation
clathrates, 219–220
inclusion, 215–216

End-functionalized polymer derivatives. *See* Poly(N-isopropylacrylamide) (PNIPAM)

Environmental changes, responses of cross-linked polymer gels, 10

Ethyl acrylate/methacrylic acid (EA/MAA) copolymer
interfacial interactions, 124–125
See also Colloidal dispersions

External stimuli, surface responsive materials, 12–13

F

Film formation
process of styrene/n-butyl acrylate (S/nBA) core and methacrylic acid (MAA) shell colloidal particles, 128f
requirements, 137
response-driven, 3
stages, 137
surfactant distribution in colloidal films, 127–129
understanding of colloidal, 123
See also Colloidal dispersions; Latex film formation

Finite element modeling for scratch damage
ABAQUS® package for finite element (FE) analysis, 167–168, 178
adaptive remeshing, 173
computational techniques, 167
contact algorithm, 172
crazing, debonding and cracking criterion, 171–172
dynamic analysis of scratch step, 169
FE mesh and plane of symmetry, 170f
geometry, element type and its boundary conditions, 170–171
load cases, 173
material law, 171
maximum envelope of equivalent plastic strain along scratch path, 177f
maximum envelope of volumetric strain along scratch path, 177f
modeling issues related to scratch, 169–173
physics of scratch deformation, 168–169
plastic yielding and crazing, 176, 178
plastic yielding criterion, 171
scratch analysis, 166–167
scratch damage process, 174, 176–178
scratch depth and width, 174, 175f
scratch research methodology, 167
static analysis of indentation and spring-back steps, 169–170
steps during scratch process, 168f
surface interaction, 172–173

Fluorescence
decay profiles of pyrene, 141f
quenching processes, 139–140
See also Latex film formation

Fluoro-surfactants
  interfacial interactions, 132, 133f
  See also Colloidal dispersions; Surfactants
Functionalized copolymers. See Reversible addition fragmentation chain transfer (RAFT)
Functional macromolecules, vs. responsive, 2–3

G

Generalized Regression Neural Networks (GRNN), 193
Geometry, scratch analysis, 170–171
Gradients, surfaces or interfaces, 2
Gradient stimuli-responsive polymer grafted layers
  gradient PAA-mix-P2VP (poly(acrylic acid)-mix-poly(2-vinylpyridine)) brush, 76, 78, 79f
  mixed gradient brushes, 76–81
  monocomponent, 72–75
  morphology upon treatment with nonselective and selective solvents, 78f
  preparation of mixed gradient brushes, 70
  preparation of polystyrene gradient brush, 69–70, 71f
  principal scheme of temperature gradient stage, 71f
  sample characterization, 70
  schematic of switching behavior of mixed polyelectrolyte (PE) brush with changing pH, 80f
  switching behavior of mixed polyelectrolyte (PE) brush, 79, 80f, 81
  switching mechanism, 69, 75, 76, 78f
  switching of water contact angle of gradient PS-mix-P2VP, 76, 77f
  synthesis of gradient brushes, 69–70
Grafting-from approach
  brushes from oxidized silicon wafers covered with initiator, 33–34
  brushes from silicon wafers with varied grafting density, 34–37
  molecular bottle brushes from silicon wafers with varied grafting density, 37, 40
  See also Stimuli responsive polymer brushes
Grafting-onto approach, stimuli-responsive ultrathin films, 30–31

H

Healing, film formation stage, 160–164
Hydrophobically modified poly(N-isopropylacrylamide). See Poly(N-isopropylacrylamide) (PNIPAM)
2-Hydroxyethyl methacrylate (HEMA)
  biocompatibility copolymers, 8
  electrografting of silylated HEMA, 93–94

I

Indentation step, static analysis of scratch, 169–170
Inimers, electrografting, 92–95
Intelligent methods. See Artificial neural networks (ANN's)
Interdiffusion process, film formation stage, 160–164
Interfaces
  gradients, 2
  poly(N-isopropylacrylamide) (PNIPAM), 109

Isoelectric point, polyelectrolyte brushes, 16–17

N-Isopropylacrylamide (NIPAAM), reversible addition fragmentation chain transfer polymerization, 48, 50, 51*f*

Isotherms. *See* Poly(*N*-isopropylacrylamide) (PNIPAM)

Isotropic phase, temperature, 13–14

L

Lactides, electrografting and ring-opening polymerization (ROP), 93–94

Lactones, electrografting and ring-opening polymerization (ROP), 93–94

Latex film formation
activation energies of void closure, minor chains, and backbone motion, 162–164
backbone activation energies vs. annealing time, 155–156
crossing density, 155
development of study techniques, 137–139
direct energy transfer (DET) methods, 138–139
emission of fluorescence, 139–140
evolution in transparency of films from high molecular (HM) and low molecular (LM) weighted particles, 156–157, 159*f*
film formation from waterborne nano-particles, 159–164
fluorescence decay profiles of film formation steps, 141*f*
fluorescence emission spectra of naphthalene and pyrene during, 145*f*
fluorescence studies, 139–149

I_{tr} (transmitted photon intensities) vs. annealing temperature, 151–152
mechanism for excited naphthalene and pyrene molecules, 148–149
molecular weight effect, 149, 151–157
particle-particle interfaces, 157
photon transmission studies, 149, 151–157
plot of back-and-forth frequencies and solubility parameters, 149, 150*f*
plot of B values vs. percentage chloroform in solvent mix, 142, 144*f*
plot of healing points, 163*f*
plot of pyrene to naphthalene intensity ratio vs. molar volume, 146–147
plot of pyrene to naphthalene intensity ratio vs. vapor exposure time, 146*f*
plot of transmitted photon intensity vs. square root of film formation time, 158*f*
plot of viscous flow activation energies vs. annealing time, 154*f*
plots of τ values vs. vapor exposure time, 142, 143*f*
poly(butyl methacrylate) (PBMA), 138
poly(methyl methacrylate) (PMMA), 138–139
poly(MMA-co-BMA) nanosized latex particles, 159–160
Prager–Tirrell (PT) model, 155
preparations for DET measurements, 144–145
requirements, 137
secondary ion mass spectroscopy (SIMS), 138
small angle neutron scattering (SANS), 137–138
stages, 137

Stern–Volmer equation for quenching kinetics, 149
vapor-induced, 140–149
void closure, healing, and interdiffusion processes, 160–164
void closure model, 152–154
See also Film formation
Liquid crystalline state, temperature and electric fields, 15–18
Liquid crystals
orientation changes with electric field, 18
surface-interfacial phenomenon, 13–14
See also Polymer dispersed liquid crystals (PDLCs)
Load cases, scratch modeling, 173
Local ionic clusters, film formation of copolymers, 129–131
Long term light exposure, polymer dispersed liquid crystals (PDLCs), 212
Lubrication, surface-interfacial phenomenon, 13–14

M

Macromolecules, responsive, 3
Macromonomers, presynthesis, 99
Magnetic fields
high-performance ceramics, 19–20
π-conjugated systems, 19
Maleimide/vinyl ether polymerization
photo-differential scanning calorimetry (photo-DSC), 202–204
See also Polymer dispersed liquid crystals (PDLCs)
Material law, scratch modeling, 171
Mathematical models. *See* Artificial neural networks (ANN's)

Mechanically assembled monolayers (MAMs), surface-interfacial phenomenon, 12
Mechanism
electrografting, 85, 87–88
electrografting of acrylonitrile, 87
See also Artificial neural networks (ANN's)
Medical devices, shape-memory effect, 5
Memory shapes, thermoplastic systems, 4–5
Metallocene polymers, nanocomposites, 15
Metal nanoparticles
formation from polyelectrolyte polymer brushes, 62–65
reversible addition fragmentation chain transfer polymers, 50, 52f
(Meth)acrylates, electrografting, 88–90
Methacrylic acid (MAA)
local ionic clusters, 129–131
See also Colloidal dispersions
3-[N-(2-Methacryloyloyethyl)-N,N-dimethylammonio]propanesulfonate (DMAPS)
chemical structure, 45f
reversible addition fragmentation chain transfer, 44
3-[2-(N-Methylacrylamido)-ethyldimethylammonio]propanesulfonate (MAEDAPS)
chemical structure, 45f
kinetics of controlled polymerization, 46f
reversible addition fragmentation chain transfer (RAFT), 44
Methyl methacrylate/n-butyl acrylate/acrylic acid (MMA/nBA/AA) copolymers, local ionic clusters, 129–131
Microporous organic polymers
molecular structures, 219f
porous solids, 218–219

Microscopic analysis, polymer dispersed liquid crystal (PDLC) films, 209–210
Modeling
scratch issues, 169–173
See also Finite element modeling for scratch damage
Molecular weight effect, latex film formation, 149, 151–157
Morphology, colloidal dispersions, 125–127

N

Nanocomposites, formation, 15
Nanoparticles
formation of metal, from polyelectrolyte polymer brushes, 62–65
loading in polymer matrices, 15
reversible addition fragmentation chain transfer polymers, 50, 52*f*
Nanoporous materials, responsiveness, 10–11
Nanostructures, formation techniques, 14–15
Naphthalene
evolution of energy transfer, 146–149
fluorescence emission spectra of, during vapor-induced latex film formation, 145*f*
See also Latex film formation
Negative Poisson ratio (NPR), thickening upon elongation, 4
Neural networks. *See* Artificial neural networks (ANN's)
N-isopropylacrylamide (NIPAAM)
reversible addition fragmentation chain transfer polymerization, 48, 50, 51*f*
See also Poly(*N*-isopropylacrylamide) (PNIPAM)

Nitroxide-mediated polymerization (NMR), sequential electrografting and, 94–95
Noncovalent interactions, self-assembly, 15
Norbornene and derivatives, electrografting and ring-opening metathesis polymerization, 93

O

Organic zeolites, porous solids, 217–218

P

Particle morphology, colloidal dispersions, 125–127
Particles
film formation process of styrene/*n*-butyl acrylate (S/nBA) core and methacrylic acid (MAA) shell colloidal, 128*f*
schematic of interactions, 131*f*
surfactant-, interactions, 127–129
See also Colloidal dispersions
pH
dependence of switching behavior, 79, 80*f*
switching behavior of polyelectrolyte brush with change of, 79, 80*f*, 81
Photochromic responses, azobenzene modified helical polymers, 18–19
Photo-differential scanning calorimetry (photo-DSC)
maleimide/vinyl ether polymerizations, 202–204
method, 197
See also Polymer dispersed liquid crystals (PDLCs)
Photon transmission studies

molecular weight effect on latex film formation, 149, 151–157
See also Latex film formation
pH-responsive materials
combining with temperature-responsive polymer, 7, 8*f*
hydrolyzed copolymer hydrogels, 9–10
poly(2-vinylpyridine)-based (P2VP), 7
reversible addition fragmentation chain transfer (RAFT), 45, 48
self-assembled porous membranes, 11
Physics, scratch deformation, 168–169
Plastic strain, maximum envelope along scratch path, 177*f*
Plastic yielding, scratch damage process, 176, 178
Plastic yielding criterion, scratch modeling, 171
Poly(acrylic acid) (PAA)
PAA-mix-poly(2-vinylpyridine) brush, 76, 78, 79*f*
pH-responsive, and temperature-responsive poly-*n*-isopropylacrylamide (PNIPAm), 7, 8*f*
polyelectrolyte diblock copolymer brushes, 58–60
Polyacrylonitrile
grafting onto common metal, 85
origin of adhesion, 88
Poly(butyl methacrylate) (PBMA), film formation measurements, 138
Poly(ε-caprolactone) (PCL)
brush of PCL chains, 94
preparation and electrografting, 99
Poly[(1-dimethoxymethylsilyl)propyl acrylate] (pDMSA), stimuli-responsive ultrathin films, 30–31
Poly(*N,N*-dimethylaminoethyl methacrylate) [pDMAEMA]
DMAEMA side chains from oxidized silicon wafers, 37–40
homopolymer brush of quaternized, 57–58
molecular weights and polydispersities, 37, 39t, 40
surface-initiated polymerization from silicon wafers, 34–37
tapping mode atomic force microscopy (AFM) of pDMAEMA grafts, 35, 36*f*
temperature responsiveness of grafts on gold, 37
Poly(*N,N*-dimethylaminoethyl methacrylate)-*b*-poly(trimethoxysilylpropyl methacrylate) [pDMAEMA-*b*-pTMSPMA], synthesis, 31, 32
Poly(dimethylsiloxane) (PDMS)
stimuli-responsive ultrathin films, 30–31
surface responsive materials, 12–13
Polyelectrolyte brushes
attenuated total reflectance–Fourier transform infrared (ATR–FTIR) spectra of diblock copolymer, 59–60, 61*f*
electrical response, 16–17
formation of metal nanoparticles from, 62–65
mixed gradient, 76, 78, 79*f*
physical properties of diblock copolymer, 60*t*
solvent treatment of diblock copolymer, 61–62
surface functionalization, 56
switching behavior with change of pH, 79, 80*f*, 81
synthesis of diblock copolymer, 58–60
synthesis of homopolymer, 57–58
See also Polymer brushes
Polyelectrolyte multilayers (PEMs), pH-triggered changes, 10
Poly(ethylacrylate) (PEA)
binary polymer films, 91–92
electrografting, 98, 99

polypyrrole nanowires through, 92, 93f
Poly(ethylene oxide) (PEO), electrografting, 98, 99
Poly(L-glutamic acid) (PLGA), pH responsive, 11
Poly(glycidyl methacrylate) (PGMA)
  anchoring layer, 70
  carboxyl terminated polystyrene, 72, 74f
Poly(2-hydroxyethyl methacrylate) (pHEMA), grafts from oxidized silicon wafers, 37–40
Poly(2-(isobutyryloxy)ethyl methacrylate), side chains from silicon surface, 37–40
Poly(N-isopropylacrylamide) (PNIPAM)
  architecture of hydrophobically modified PNIPAM (HM-PNIPAM), 110, 111f
  chemical structure of derivatives, 111f
  compression-expansion cycles of monolayers of C_{18}-PNIPAM-C_{18}, 116f
  compression-expansion cycles of monolayers of C_{18}-PNIPAM-C_3, 115f
  compression-expansion cycles of monolayers of PNIPAM-F, PNIPAM-H, and C_4-PNIPAM-C_4, 113f, 114f
  compression isotherms of monolayers of PNIPAM-F, PNIPAM-H, and C_4-PNIPAM-C_3, 112f
  conceptual representations of end-functionalized PNIPAM derivatives at air/water interface, 118f
  conceptual representations of graft PNIPAM derivatives at air/water interface, 117f
  differences under high compression conditions, 117–118
  effect of temperature on compression isotherms, 118–119
  hydrogen bonding, 109
  hydrophobically modified (HM-PNIPAM), 109–110
  interfacial properties in water, 109
  isotherms of end-functionalized PNIPAM derivatives, 114–119
  isotherms of graft-PNIPAM derivatives, 110–114
  overlap of expansion and compression isotherms, 115–116
  preparation of HM-PNIPAM, 109
  self-aggregation, 108
  stability of PNIPAM Langmuir films, 119
  surface properties, 108–109
Polymer brushes
  block copolymer brushes on flat surfaces, 33–40
  definition, 29
  preparation, 17
  surface modification, 56
  See also Polyelectrolyte brushes; Stimuli responsive polymer brushes
Polymer chain topology. See Stimuli responsive polymer brushes
Polymer dispersed liquid crystals (PDLCs)
  commercial bis-maleimide (Q-BOND), 197, 198
  divinyl ether VEctomer samples, 197, 198
  electro-optic behavior of thiol-ene based PDLCs, 209
  electro-optic measurements, 200, 201f
  electro-optic responses of, 208f
  electro-optic switching data for maleimide/divinyl ether PDLCs, 207t

exotherm maximum (H_{max}) vs. mole percent Q-BOND in mixtures of Q-BOND, VEctomer 4020, and VEctomer 2020, 206f
experimental, 197–200
long term light exposure evaluation, 212
materials and sample preparation, 197
microscopic analysis of PDLC films, 209–210
photodifferential calorimetry (photo-DSC) method, 197
photo-DSC exotherms for maleimide/vinyl ether polymerization, 202–204
photo-DSC exotherms for mixtures of Q-BOND and VEctomer 4020, 204f
photo-DSC for Q-BOND and VEctomer 4020 with no added photoinitiator, 201f, 203f
plot of H_{max} vs. mole percent Q-BOND in mixtures of Q-BOND and VEctomer 4020, 205f
polymerization-induced phase separation, 196
pre-polymer mixture formulations, 202t
refractive index measurements, 200
scanning electron microscopy (SEM), 200
schematic of PDLC cell, 200f
SEMs of PDLC films, 210f, 211f
structures of monomers and LCs comprising E7, 198, 199
vinyl ether VEctomer samples, 197, 198
Polymeric films, response-driven, 3
Polymeric materials, environmental stimuli functions, 2
Polymers, stress-responses, 3–6
Poly(methyl acrylate) (PMA), polyelectrolyte diblock copolymer brushes, 58–60
Poly(methyl methacrylate) (PMMA) film formation measurements, 138
See also Latex film formation
Poly(pyrrole) (PPy)
electroactivity, 91–92
nanowires, 92, 93f
Poly(styrene)
gradient grafting of carboxyl-terminated, 72, 73f
mixed brush of poly(2-vinylpyridine) (P2VP) and, 76, 77f
polyelectrolyte diblock copolymer brushes, 58–60
stimuli-responsive ultrathin films, 30–31
Poly(tetrafluoroethylene) (PTFE), morphological changes with elongation, 4
Polyvinylamine (p-VAm), water-soluble synthetic polymers, 9
Poly(vinylpyridine) (PVP). *See* Vinylpyridine (VP)
Poly(2-vinylpyridine) (P2VP), pH-responsive, 7
Poly(4-vinylpyridine)-based (P4VP) latex, acid-swellable microgels, 7
Porous solids
classification, 216
microporoud organic polymers, 218–219
organic zeolites and biozeolites, 217–218
zeolites and related materials, 216–217
See also Supramolecular solids
Prager–Tirrell's crossing density, model, 142, 155
Pressure sensitive adhesives (PSA), surface-interfacial phenomenon, 11
Probabilistic Neural Networks (PNN), 193

Pyrene
  evolution of energy transfer, 146–149
  fluorescence decay profiles, 141f
  fluorescence emission spectra of, during vapor-induced latex film formation, 145f
  labeled polymer particles, 140
  See also Latex film formation

Q

Quenching processes
  fluorescence, 139–140
  low efficiency, 142

R

Radical addition fragmentation chain transfer, hydrophobically modified poly(N-isopropylacrylamide), 109
Random copolymer brush, electrografting and nitroxide-mediated polymerization, 94–95
Reactive polymers
  direct grafting of preformed, 100
  electrografting, 97–100
Refractive index
  changes of bismaleimide/divinyl ether based polymer dispersed liquid crystals (PDLCs), 204, 207
  measurements, 200
  PDLC films before and after polymerization, 207t
Release coatings, interface, 11
Re-mending ability, polymeric materials, 5
Responsive macromolecules
  electrical and magnetic fields, 15–20
  functional vs., 2–3

Reversible addition-fragmentation chain transfer (RAFT)
  application of RAFT-generated polymers, 50, 52
  characterization techniques and instrumentation, 52
  combining pH- and temperature-responsive polymers, 7, 8f
  experimental, 52
  methodology, 44
  monomer and polymer syntheses, 52
  synthesis of electrolyte-responsive polymers, 44, 45f
  synthesis of pH-responsive polymers, 45, 47f, 48, 49f
  synthesis of temperature-responsive polymers, 48, 50, 51f
Ring-opening metathesis polymerization (ROMP), sequential electrografting and, 93
Ring-opening polymerization (ROP)
  poly(ε-caprolactone) (PCL) preparation, 99
  sequential electrografting and, 93–94

S

Scanning electron microscopy (SEM)
  method, 200
  microscopic analysis of PDLC films, 209–210
  polymer dispersed liquid crystals (PDLCs), 210f, 211f
Scratch damage
  analysis, 166–167
  depth and width, 174, 175f
  equivalent plastic strain along scratch path, 177f
  mechanical deformation, 166
  modeling issues, 169–173
  physics of scratch deformation, 168–169

process, 174, 176–178
volumetric strain along scratch path, 177f
See also Finite element modeling for scratch damage
Self-assembled liquid crystals, ion-responsive, 16
Self-assembled monolayers (SAMs), surface-interfacial phenomena, 11–12
Self-assembly
2-hydroxyethyl methacrylate (HEMA) copolymers, 8
noncovalent interactions, 15
poly(N,N-dimethylaminoethyl methacrylate)-b-poly(trimethoxysilylpropyl methacrylate) [pDMAEMA-b-pTMSPMA], 32
poly(N-isopropylacrylamide) (PNIPAM), 108
sodium dodecyl sulfate (SDS), 132, 133f
See also Poly(N-isopropylacrylamide) (PNIPAM)
Self-healing polymer coatings, development, 5
Shape memory materials, thermally induced effect, 4–5
Shear stresses, solution viscosity and, 5–6
Silicon wafers
brushes from, with varied grafting density, 34–37
brushes from oxidized, covered with initiator, 33–34
molecular bottle brushes from, with varied grafting density, 37, 40
See also Stimuli responsive polymer brushes
Silver nanoparticles, formation from polyelectrolyte polymer brushes, 63–65
Smart sorbent

dynamic sorbents, 221–224
processing scheme, 223f
See also Supramolecular solids
Smectic phase, temperature, 13–14
Sodium dioctyl sulfosuccinate (SDOSS)
interactions with ethyl acrylate/methacrylic (EA/MAA) copolymer dispersions, 124–125
mobility during film formation, 125–127
surfactant-particle interactions, 127–129
vibrational dependencies, 124f
See also Colloidal dispersions
Sodium dodecyl sulfate (SDS)
migration to interfaces, 125
mobility during film formation, 125–126
surface self-assembly in presence of other surfactants, 132, 133f
See also Colloidal dispersions
Solids. *See* Supramolecular solids
Solution viscosity, applied shear stresses, 5–6
Solvents
electrografting, 88–90
gradient grafted polymers, 74, 75
gradient mixed brush morphology with nonselective and selective, 76, 78f
Solvent treatment, polyelectrolyte diblock copolymer brushes, 61–62
Sorption, inclusion, 215
Spectroscopic techniques, adhesion, 88
Spring-back step, static analysis of scratch, 169–170
Static analyses, indentation and spring-back steps, 169–170
Stern–Volmer equation, quenching, 140, 142
Stimuli-response mechanisms. *See* Artificial neural networks (ANN's)
Stimuli responsive polymer brushes

AFM (atomic force microscopy) images of poly(2-hydroxyethyl methacrylate) (pHEMA) and polymer brush, 40
AFM of Si–OH vs. Si–H substrates, 36*f*
attaching initiator to Si–H substrates, 35
attachment of atom transfer radical polymerization (ATRP) initiators and dummy initiators to surface of silicon wafers, 34
attachment of 1-(chlorodimethylsilyl)propyl 2-bromoisobutyrate as ATRP initiator to silicon wafer, 33
block copolymer brushes on flat surfaces, 33–40
brushes from oxidized silicon wafers covered with initiator, 33–34
brushes from silicon wafers with varied grafting density, 34–37
copolymer synthesis of poly(*N,N*-dimethylaminoethyl methacrylate)-*b*-poly(trimethoxysilylpropyl methacrylate) [pDMAEMA-*b*-pTMSPMA], 32
grafting-from approach, 33–40
grafting-onto approach of functional ABC triblock copolymer, 30–31, 32*f*
homogeneity of surface coverage, 35
molecular bottle brushes from silicon wafers with varied grafting density, 37, 40
molecular weights and polydispersities of pDMAEMA, 39*t*, 40
pHEMA, 37, 39*f*
self-assembly of pDMAEMA-*b*-pTMSPMA to glass slide surface, 32
synthesis of grafting-onto approach with pDMAEMA-*b*-pTMSPMA, 32
synthesis of poly[(2-(isobutyryloxy)ethyl methacrylate)-*g*-(*N,N*-dimethylaminoethyl methacrylate)] brush attached to silicon wafers, 38
tapping force AFM images of pDMAEMA grafts, 36*f*
tapping mode AFM height images of pDMS-*b*-pSt-*b*-pDMSA brushes, 31*f*
water contact angles and modified silicon wafers, 33–34
Stimuli-responsive polymers and films
applications, 2
behavior of colloidal dispersions, 123
crystalline phase orientation and elongation, 3–4
functional vs. responsive, 2–3
morphology and elongation, 3–4
multi-dimensional arrays, 3
pH-responsive gels, 7, 8*f*
responses to electric and magnetic fields, 15–20
self-healing polymeric coatings, 5
shape memory materials, 4–5
solution viscosity and shear stresses, 5–6
stress-responses, 3–6
surface-interfacial phenomena, 10–14
surfaces or interfaces, 2
swelling-shrinking behavior of water-soluble polymers, 9–10
temperature and pH, 6–10
temperature dependent self-assembly, 8
See also Colloidal dispersions; Gradient stimuli-responsive polymer grafted layers

Styrene/ethyl hexyl
  acrylate/methacrylic acid
  (Sty/EHA/MAA), crosslinking and
  particle morphology, 125–127
Supramolecular solids
  bonding hierarchy, 215
  clathrates, 219–221
  dynamic sorbents, 221–224
  inclusion behaviors, 215–216
  microporous organic polymers,
    218–219
  molecular structures of
    microporous organic polymers,
    219f
  organic zeolites and biozeolites,
    217–218
  porous solids, 216–219
  processing scheme of smart
    sorbent, 223f
  range of materials, 215
  sorption, clathration, and dynamic
    sorption, 221f
  zeolites and related materials, 216–
    217
Surface interaction, scratch modeling,
  172–173
Surface-interfacial phenomena
  adhesive and lubricant coatings,
    13–14
  atomic force microscopy (AFM),
    10–11
  biological and biomedical
    materials, 10
  externally responsive materials,
    12–13
  mechanically assembled
    monolayers (MAMs), 12
  nanoporous membranes, 10–11
  pressure sensitive adhesives (PSA),
    11
  self-assembled monolayers
    (SAMs), 11–12
Surface modification, polymer
  brushes, 29, 56
Surfaces
  gradients, 2
  modifications for tailoring
    properties, 29
  poly(N-isopropylacrylamide)
    (PNIPAM), 108–109
Surfactants
  interactions of fluoro-surfactants,
    132, 133f
  interactions with particles, 127–
    129
  mobility during film formation,
    125–127
  See also Colloidal dispersions
Swelling-shrinking behavior,
  hydrolyzed copolymer hydrogels,
  9
Switching behavior
  gradient grafted polymers, 69, 75
  gradient poly(acrylic acid)-mix-
    poly(vinylpyridine) [PAA-mix-
    PVP], 76, 78, 79f
  mixed polyelectrolyte (PE) brush
    varying pH, 79, 80f, 81
  tuning value of wetting gradient for
    PAA-mix-PVP brush by pH, 79,
    80f
Synthetic polymers, surface
  responsive materials, 12–13

T

Tackiness, surface-interfacial
  phenomenon, 13–14
Temperature
  annealing film samples at elevated,
    154–156
  compression isotherms, 118–119
  gradient grafted polymer layers, 72
  gradient on stage surface, 71f
  liquid crystal transitions, 13–14
  polystyrene brushes, 71f
Temperature-responsive systems
  combining with pH-responsive
    poly(acrylic acid) (PAA), 7, 8f

end-functionalized poly(*N*-isopropylacrylamide) (PNIPAM) derivatives, 114–119
graft poly(*N*-isopropylacrylamide) (graft-PNIPAM) derivatives, 110–114
hydrolyzed copolymer hydrogels, 9–10
2-hydroxyethyl methacrylate (HEMA) copolymers, 8
poly(*N,N*-dimethylaminoethyl methacrylate) (pDMAEMA) grafts on gold, 37
poly(*N*-isopropylacrylamide) (PNIPAm), 6, 7*f*
reversible addition fragmentation chain transfer (RAFT) polymerization, 48, 50, 51*f*
See also Poly(*N*-isopropylacrylamide) (PNIPAM)
Thiol-ene based polymer dispersed liquid crystals, electro-optic behavior, 209
Transmitted photon intensities, low and high molecular weighted latex films, 151–152
Triblock copolymers, stimuli-responsive ultrathin films, 30–31

U

Ultrathin films, stimuli responsive, by grafting-onto, 30–31

V

Vapor-induced film formation. *See* Latex film formation

n-Vinylformamide (NVF), thermo- and pH-responsive copolymers, 9
n-Vinylisobutyramide (NVIBA), thermo- and pH-responsive copolymers, 9
Vinylpyridine (VP)
copolymerization of 2-VP and 4-VP, 45, 47*f*
homopolymerizations of 2-VP and 4-VP, 45, 47*f*
kinetics of controlled polymerization of 2-VP, 47*f*
mixed brush of poly(2-VP) (P2VP) and polystyrene (PS), 76, 77*f*
poly(acrylic acid)-mix-poly(2-VP) mixed gradient brush, 76, 78, 79*f*
switching behavior of mixed polyelectrolyte (PE) brush with varying pH, 79, 80*f*, 81
Void closure, film formation stage, 160–164
Void closure model, latex film formation, 152–154
Volumetric strain, maximum envelope along scratch path, 177*f*

W

Waterborne nano-particles, film formation, 159–164
Water contact angles, polymer brushes on modified silicon wafers, 33–34
Water-soluble polymers, swelling-shrinking with temperature and pH, 9–10
Wettability, surface-interfacial phenomenon, 13–14
Wetting behavior, gradient grafted polymer layers, 72–74

X

X-ray photoelectron spectroscopy (XPS), metal content in diblock copolymer brushes, 65

Z

Zeolites, porous solids, 216–217